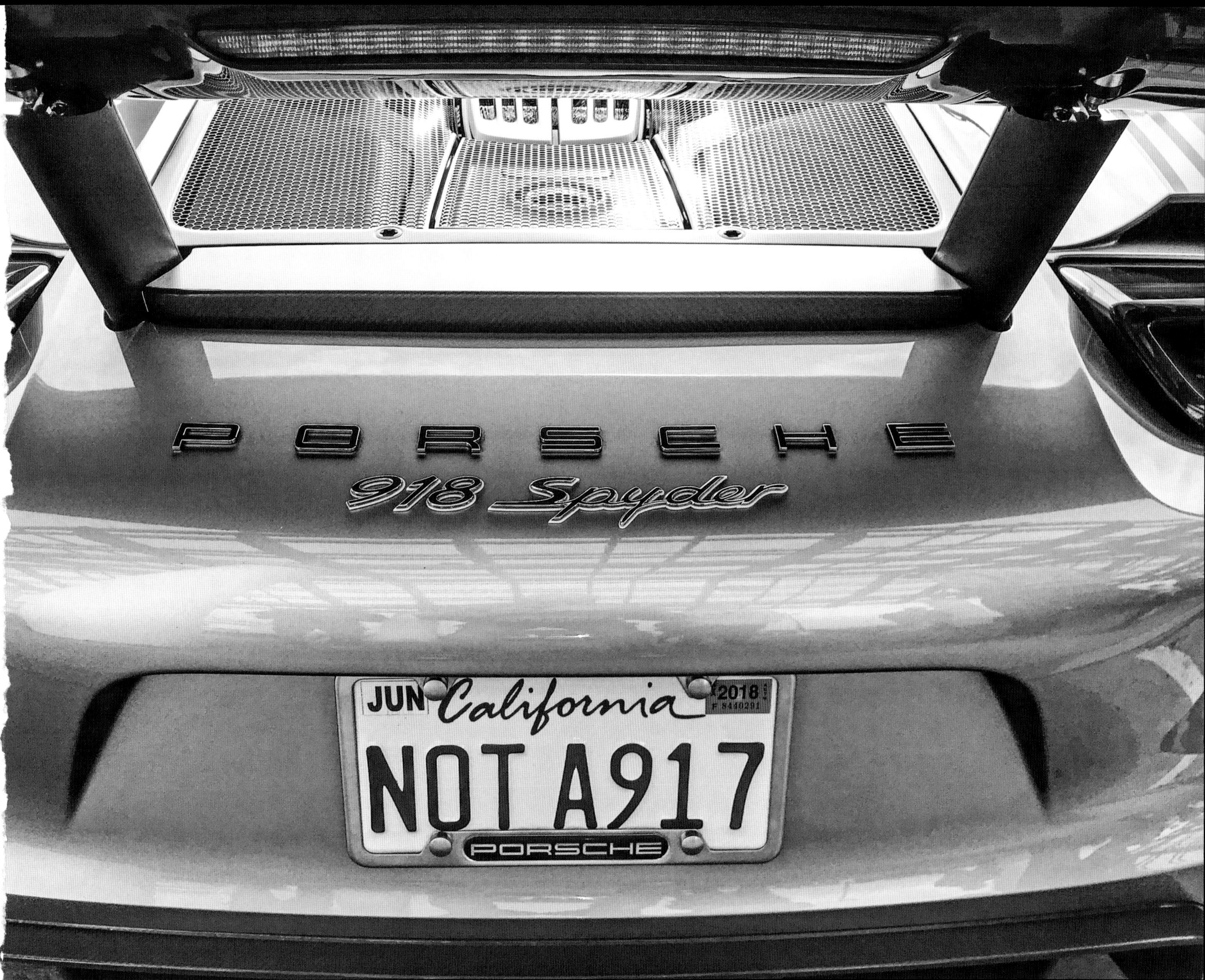
PORSCHE
918 Spyder
JUN
California
2018
NOT A917
PORSCHE

Brimming with creative inspiration, how-to projects, and useful information to enrich your everyday life, Quarto Knows is a favorite destination for those pursuing their interests and passions. Visit our site and dig deeper with our books into your area of interest: Quarto Creates, Quarto Cooks, Quarto Homes, Quarto Lives, Quarto Drives, Quarto Explores, Quarto Gifts, or Quarto Kids.

First published in 2018 by Motorbooks, an imprint of The Quarto Group, 100 Cummings Center, Suite 265-D, Beverly, MA 01915, USA. T (978) 282-9590 F (978) 283-2742 www.QuartoKnows.com

Motorbooks titles are also available at discount for retail, wholesale, promotional, and bulk purchase. For details, contact the Special Sales Manager by email at specialsales@quarto.com or by mail at The Quarto Group, Attn: Special Sales Manager, 100 Cummings Center, Suite 265-D, Beverly, MA 01915, USA.

10 9 8 7 6

ISBN: 978-0-7603-6334-8

Digital edition published in 2018
eISBN: 978-0-7603-6335-5

Library of Congress Cataloging-in-Publication Data

Names: Lamm, John, author.
Title: Supercar revolution : the fastest cars of all time / by John Lamm.
Description: Minneapolis, Minnesota : Motorbooks, 2018. | Includes index.
Identifiers: LCCN 2018022848| ISBN 9780760363348
(paper over board + jacket)
ISBN 9780760363355 (ebook)
Subjects: LCSH: Sports cars. | Automobiles, Racing.
Classification: LCC TL236 .L3529 2018 | DDC 629.222--dc23
LC record available at https://lccn.loc.gov/2018022848

Acquiring Editor: Darwin Holmstrom
Project Manager: Jordan Wiklund
Art Director: Brad Springer
Cover Designer: Jay Smith - Juicebox Designs
Layout: Simon Larkin

On the front cover: Ferrari LaFerrari
On the back cover: McLaren P1
On the front flap: Aston Martin DB11
On the back flap: Pagani Huayra
On the endpapers: Dashboards of the McLaren F1 (front) and Ford GT 2005 (rear)
On the frontis: Porsche 918
On the title page: Lamborghini Countach LP400

Printed in China

SUPERCAR
REVOLUTION

THE FASTEST CARS OF ALL TIME

JOHN LAMM

motorbooks

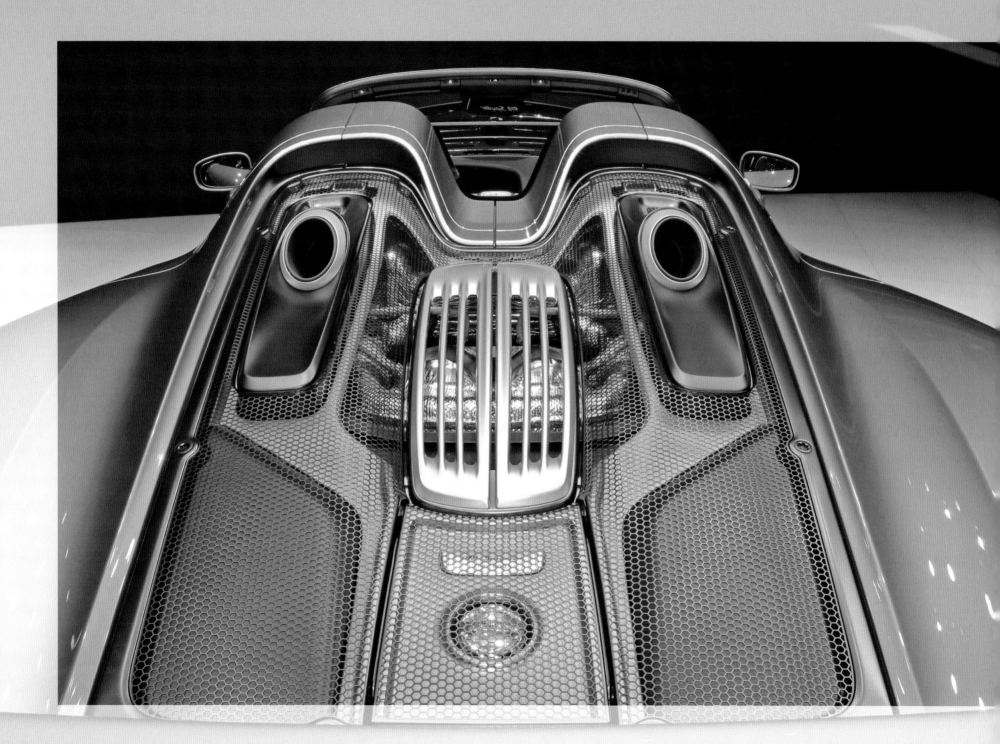

Contents

Foreword   6

Acknowledgments   6

Introduction   7

SECTION I:
THE FIRST WAVE: 1967–1978

Lamborghini Miura & Ferrari Daytona 1967   10

Maserati Bora 1971   22

Ferrari Boxer 1973   28

Lamborghini Countach 1974   38

Porsche Turbos 1975   48

BMW M1 1978   56

SECTION II:
THE GROUP B CONNECTION: 1983–1991

Ferrari 288 GTO 1983   64

Ferrari Testarossa 1984   70

Ferrari F40 1987   76

Jaguar XJ220 1988   86

Porsche 959 1988   94

Lamborghini Diablo 1990   102

Bugatti EB110 1991   110

The Vector 1991   114

Honda NSX 1991   120

SECTION III:
THE MODERN SUPERCAR ERA: 1992–PRESENT

McLaren F1 1992   126

Ferrari F50 1996   138

Pagani Zonda 1999   144

Saleen S7 2001   148

Ferrari Enzo 2002   154

Maserati MC12 2004   162

Porsche Carrera GT 2004   166

Ford GT 2005   176

Bugatti Veyron 16.4 2005   184

Mercedes-Benz SLS AMG 2010   192

Pagani Huayra 2012   196

Porsche 918 2013   200

McLaren P1 2013   204

Lamborghini Huracán 2014   208

Ferrari 488 GTB 2016   214

Bugatti Chiron 2016   218

Aston Martin DB11 2016   222

Ford GT 2017   226

Lexus LC 500 2017   230

Corvette ZR1 2019   234

Index   238

Foreword

The vehicles of *Supercar Revolution* are near and dear to my heart. I have been involved with them since the 1970s, and I have delighted in driving them and writing about them for *Road & Track* during these many years. I have also very much enjoyed working with John Lamm for more than thirty years. He is a special talent in the world of automotive journalism because he is both an excellent writer and among the most highly respected automotive photographers in the world.

John and I have traveled the world together on many assignments, and his knowledge and understanding of these exotic cars is without parallel. Add to that his enthusiasm for going anywhere to get the job done, his love for his craft, and his determination to do it right, and the result is this remarkable book you hold in your hands.

Thos L. Bryant
Former Editor-in-Chief
Road & Track Magazine

Acknowledgments

This book covers the work of many decades, so chances are I've forgotten someone who helped me. To them, my deepest apologies. Thanks to Bill Baker, Chris Bangle, Richard Baron, Michael Baumann, Toscan Bennett, Klaus Bischof, Thos L. Bryant, Luca Dal Monte, Ian Callum, Harry Calton, Anne-Helene Casse, John Clinard, Geoff Day, Pietro DiFranchi, Bill Donnelly, Jeffrey Ehoodin, Leonardo Fioravanti, Sergio Fontana, Paul Frere, Marcello Gandini, Jack Gerkin, Antonio Ghini, Joe Grecco, Bernd Harling, Phil Hill, Dominik Hoberg, Patrick Hong, Georges Keller, Davide Kluzer, Doug Kott, Ralph Lauren, Richard Losee, Rob Mitchell, Rob Moran, Sandro Munari, Gordon Murray, Ken Okuyama, Horace Pagani, Chuck Queener, Mark Reinwald, Allan Rosenberg, Alois Ruf, Joe Sackey, Steve Saleen, Michael Schimpke, Murray Smith, Ken Sodowsky, Frank Stephenson, and Bert Swift.

A special thank you to Jay Leno for his personal comments about many of the cars. Thanks also to Tim Parker for his help in developing the concept for this book.

Introduction

We were headed downhill, so when the throttle fell open we quickly gained speed and things got tense. The brakes were questionable and the wheels were making a funny noise. Thank goodness world driving champion Phil Hill was behind the wheel, or we might have ended up . . . I hate to think about the last part of that sentence. Velocity can be scary stuff when it gets out of hand.

We were in an 1893 Benz Victoria that topped out around 30 miles per hour, which was also scary because velocity is relative. Relative velocity is also relatively inexpensive in any era, but if you want to test the upper limits, you need to slip into the somewhat rarified air of supercars or, as some prefer, exotic cars.

There have been super and exotic automobiles from the beginning. In its time, Gottlieb Daimler's 1886 huffing-and-puffing single-cylinder Patent Motorwagen was exotic, and it must have been super to get from Point A to Point B without looking at the backside of a horse. Automotive history is riddled with other examples of great machines such as Stutz Bearcats, Mercedes-Benz SSKs, Bugatti Type 57s, Duesenberg SJs, 8C 2900B Alfa Romeos, and generations of hot rods that will get your heart pumping faster than a gallon of espresso.

After World War II, the line never lost speed, with Chrysler 300s, Mercedes-Benz 300SLs, Maseratis, Jaguars, and—of course—Ferraris.

It can be argued that you can't get any more exotic than the Ferrari Superfasts. Very exclusive, they were built as one-offs or in small series. The very first one in 1956 was wrapped in Pininfarina bodywork that would stop you in your tracks. Its drivetrain was straight out of a Ferrari 410 Sport Spyder race car. Talk about velocity . . .

Some would argue that this book should begin with the start of series Ferrari production, with the 250 GT Lussos, the 275/330 GTCs, and 275 GTBs. Across town in Modena, Italy, Maserati was crafting the beautiful Giorgetto Giugiaro–designed Ghibli. Others would suggest that we throw in the Cobra Daytona coupe and Corvette Grand Sport.

Yet somehow the modern era of super and exotic cars seems to flow from one day in March 1966, when the Lamborghini Miura was unveiled at the Geneva Auto Show. On this day, everyone's attention was diverted from Ferrari to consider the most evocative automotive shape ever penned . . . and by the fact that a 320-horsepower engine sat sideways behind the driver . . . and when you'd gone through all five gears, you were at 163 miles per hour.

Ferrari was shaken. There was an honest alternative to the Prancing Horse—the charging bulls of Lamborghini.

The modern supercar and exotic car wars were on. And that's where *Supercar Revolution* begins.

SECTION I:

The First Wave: 1967–1978

Race cars of the late 1960s, which were shaped more for function than for beauty, influenced Italian Marcello Gandini's design of the Countach. Gandini said of his design that he wanted "people to be astonished when they saw the car." They were. Before his work on the Countach, however, Gandini penned the Alfa Romeo Carabo, first seen at the Paris Auto Show in October 1968. It is arguably the most influential exotic show car of the modern era.

155-190^{mph}

Lamborghini's Miura (left) and Ferrari's Daytona mark the beginning of the modern supercar era. Lamborghini took the engineering lead by giving the Miura a mid-mounted transverse V-12 engine.

Lamborghini Miura &
Ferrari Daytona 1967

163^{mph} & 173^{mph}

As the crow flies, Enzo "Commendatore" Ferrari and Ferruccio Lamborghini worked only 17 miles apart, but their thinking was on different planets.

Enzo Ferrari was conservative, so his 365 GTB/4 Daytona had a front engine and rear drive. Ferrari always said the horse belongs in front of the chariot. The car drove like a traditional Ferrari, with wonderful mechanical noises from the V-12 ahead of you and a tall shift lever in a metal gate to your right, and when you punched the throttle, you were pulled along by powerful forces. With speed, you sensed you were in the last car of the roller coaster.

Ferruccio Lamborghini was a former Ferrari owner who badly wanted to trump the Commendatore. So he approved the nontraditional Miura with a midmounted engine and a low nose that offered a widescreen view of the road. The Lamborghini's snarling V-12, sideways in back, seemed to propel you ahead. You were in the roller coaster's front car.

Such different racehorses come from basically the same breeding ground.

Great cultural changes influenced automobile design in the early 1960s, and different countries seemed to lead the way in various sectors. The English headed the midengine movement in race cars. Porsche did the best job of cheating the wind. American tire companies seemed to make their products wider each week. And—with a few notable exceptions, such as Jaguar's E-Type—the Italians held the high ground in exterior design.

This latter trend wasn't a new one, and it was surprisingly localized. Perhaps because their grandfathers were famous for shaping metal pots, suits of armor, and gun barrels, young men in the area around the northern Italian city of Torino seemed to have a way with forming automotive bodywork. Giovanni Agnelli had established Fiat in Torino, giving the locals plenty of metal to shape. It was no accident that automotive design houses Pininfarina, Bertone, Ghia, and Italdesign are a short drive from each other in greater Torino.

Let's narrow it down even more. Between January and August 1938, three male *bambini*, Leonardo Fioravanti, Marcello Gandini, and Giorgetto Giugiaro, were born near Torino. Combined, this trio of designers was responsible for—among many others—the Lamborghini Countach; the Ferrari Dino, Boxer, and 308; the Lotus Esprit; the Detomaso Mangusta; the DeLorean; the BMW M1; the Alfa Romeo Carabo; the Lancis Stratos; the Lamborghini Miura; and the Ferrari Daytona.

Now jump ahead to mid-1964, as Lamborghini engineers Giampaolo Dallara (who makes Indy Racing League chassis today) and Paolo Stanzani conspired after hours with ace mechanic and test driver Bob Wallace to create a midengine exotic car. Though ever more common in race cars, this powertrain layout had not yet been used in a series production sports machine—yet Ferruccio Lamborghini approved the project.

A chassis was ready for the Turin Motor Show in late 1965 and labeled the P400 (*posteriore* 4.0-liter engine). Based on a boxed central frame of sheet steel, with extensions front and back for the drivetrain and upper and lower A-arm suspensions, the platform looked simple and elegant with its lightning holes. The company's 4.0-liter, twin-cam V-12 was mounted transversely behind the driver, the five-speed gearbox combined in a cast housing with the crankcase and their mutual lubricating oil. While considered a fascinating and promising study, the chassis was missing two crucial elements: bodywork and a name.

Designer Marcello Gandini was only in his twenties when his boss, Nuccio Bertone of Carrozzeria Bertone, brought him the challenge of the new Lamborghini. He remembers, "After the P400 was presented at the Turin Motor Show in 1965, Mr. Lamborghini, with engineers Dallara and Stanzani, contacted the Carrozzeria Bertone for a body for this chassis, which had a racing car definition. At first glance, it was clear that we could make a very sporty car out of it."

What were his instructions? "There was no brief at the time, but we all agreed that the body had to underline—and not hide—the evident and aggressive muscles of the frame."

Gandini, who later designed the body for Lamborghini's Countach, calls the Miura "a good synthesis of the tradition of the famous sporty car from the 1950s and 1960s, but interpreted in a modern way. And that's why everybody liked it. This car was interesting and new enough to amaze, but with soft shapes that gave to the eye a certain pleasure." He adds, "This wasn't the case with the Countach, which was more bothering, at least at a first glance, in its unusual shapes."

When the Geneva Auto Show opened in March 1966, the P400 had its exotic bodywork, including retracting headlamps with eyebrows and flaming orange paintwork, and a name—Miura—from a breed of Spanish fighting bulls. The reviews were excellent, and the car established Lamborghini once and for all as a potential rival for Ferrari.

There has been some controversy over the part Giorgetto Giugiaro played in the design of the Miura. The first person to dispel such rumors is Giugiaro himself, who quickly points out that he left the company in

No photo can capture the sound of the Miura's V-12 hunkered down under those amazing rear window louvers, nor how large the car appears in person; the overall length of the Lamborghini is 171.6 inches, and the width is 69.3 inches, but the Miura's height is a mere 41.5 inches.

The Miura first appeared as a bare chassis at the Turin Auto Show in late 1965. Engineers Giampaolo Dallara and Paolo Stanzani worked after hours with Bob Wallace to create the spectacular midengine chassis.

Designer Marcello Gandini calls his Miura "a good synthesis of the tradition of the famous sporty car from the 1950s and 1960s, but interpreted in a modern way."

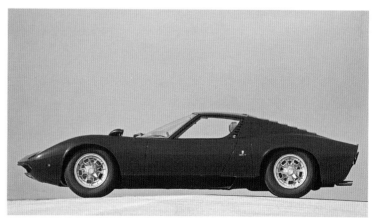

November 1965 for Ghia. Gandini took his place, and the Miura was unveiled at Geneva the following March.

Giugiaro had, however, spent his six years at Bertone working with its team, creating for the design house a midengine style that has been part of the Miura's legacy. He recalls that the strong economy of the 1960–1963 period prompted many new automobiles, and that several automakers wanted to be in the race to build the first midengine production car.

Giotto Bizzarrini was a likely candidate but lacked funds. Alessandro De Tomaso also wanted Giugiaro, by way of Bertone, to create a midengine model, but money proved to be a problem there too. Nuccio Bertone, being a good manager, realized that the logical automaker to build the first midengine car was Lamborghini.

For all the Miura's power and potential, the factory never raced it. Some critics argue that Ferruccio Lamborghini's refusal to go racing kept his company from challenging cross-town rival Ferrari.

The staff of *Road & Track* admitted the Daytona was not the most exotic car in the world, but declared it to be " . . . the best sports car in the world. Or the best GT. Take your choice, it's both." When new, a Daytona cost $19,500, a small percentage of what one is worth today. This Daytona has the traditional Borrani wire wheels, a feature of the earliest cars. Later cars featured alloy wheels—designed for Ferrari by Cromodora—that became so famous the shape was used on other cars, where they are simply called "Daytona wheels."

Before Giugiaro's departure, Nuccio Bertone asked him to design a midengine car. He was given the car's dimensions, the packaging of the engine, the pedals, and the steering, but not a name, though he suspected it might be Bizzarrini. He points out that the transition to a midengine sports car layout was more of a technical problem for engineers rather than designers. And, like several men involved in the first midengine designs, Giugiaro reveals, "The Ford GT40 [the early racing version], with respect to the development of sports cars of the time, was very important to us."

What did Giugiaro think when he first saw the Miura? "It was an impressive work, a brand-new kind of sports car." He explains, "You can say many things about the Miura; you can like it for some things, you can dislike it, but the greatest achievement is that he found a way to put sporty racing cars on the street in a very successful way and for the first time."

"The Miura was the first to have the driver's position put so far forward," Giugiaro adds. This presented a new ratio between fundamental dimensions, such as front and rear overhang and the placement of a cockpit and engine bay, so "optically it was a real breakthrough," Giugiaro

says. The Lamborghini also provided a psychological impact, according to Giugiaro. "The idea of a long hood was associated with prestige and luxury, so taking the engine back and changing this proportion while still having a serious, racy car really was a breakthrough."

After Giugiaro went to Ghia, Alessandro De Tomaso finally got his Giugiaro-designed midengine car, the Mangusta. Giugiaro points out that because he had already given his ideas to Bertone for the earlier midengine car, he was constrained when it came time to design the Mangusta. He wanted it to be different from the rounded shapes he had done at Bertone, and he made an "absolute effort"—laying the windscreen back, moving the B-pillar—to visually separate the two cars.

Where the Lamborghini Miura pointed ahead with both its chassis layout and exterior design, the Ferrari Daytona was more conservative but also designed in such a way as to separate it from its predecessors.

The Daytona's is a simpler story than the Miura's. In 1965, Leonardo Fioravanti was a young designer at Pinifarina with a very active imagination and a lot of independence to do what he liked, when "one day there arrived at Pininfarina in Grugliasco, a mechanical chassis complete with wheels, engine, steering wheel, and fuel tank. By chance, I saw it immediately after it was discharged from the truck," Fioravanti says. "I was shocked, thinking that until this moment we were completely wrong."

Fioravanti can't recall the project he was working on at the time, but he immediately put it aside. "I was driven by an extreme desire to design a new Ferrari. So I worked one week continuously. My wife only saw me late in the evening. Finally, I finished my work . . . three to four views, some perspectives."

Presented with the drawings, Sergio Pininfarina was a bit surprised; the designer recalls the conversation: "'A Ferrari? But there is no plan for a new Ferrari.' I explained why I did it, and finally he looked at the design and was very impressed. Soon after, Mr. Pininfarina met Mr. Ferrari and told him, 'We have a possibility to propose to you—a new GTB.'"

Enzo Ferrari was interested, possibly because his current grand touring model, the 275 GTB/4, wasn't selling that well. In this decade, when a very nice GTB sells for $400,000, that seems so odd, but those were the times.

Fioravanti explains it from his perspective: "To me the 275 GTB/4 wasn't the best Ferrari. Now it is considered one of the best, but to me it was a bit too old, not with the best proportion to the side. It was strong evolution on all the themes from Ferrari's past."

Ferrari gave permission to start on the new car using the GTB chassis, which meant a steel tube frame, independent upper and lower A-arm suspension at both ends, and that most sublime of powerplants, the Ferrari V-12, now in 365 twin-cam form with 4.4 liters and 405 brake horsepower. While that classic layout meant the new car would be technically less interesting than the Miura, it brought with it a heritage and almost mystical aura that Lamborghinis had never had, thanks to Ferruccio Lamborghini's anti-racing policy.

As it turned out, the most difficult part of the Daytona design was the front, because the GTB chassis proved to be too narrow for its length. Seen on a 1:1 model, the original design looked quite nice, but as Fioravanti recalls, "the main criticism was the front. Very disappointing. Too thin. Too high from the ground. In my view, too old."

On one of his rare trips outside of Maranello, Enzo Ferrari visited Pininfarina in Torino to see the full-size Daytona model. He liked

In its day, a Daytona would get to 60 miles per hour in 5.4–5.9 seconds, with a top speed around 175 miles per hour, depending on the car's specifications. Ferrari didn't officially name the car Daytona, but the nickname was started to commemorate the Ferrari 330 P4's 1-2-3 win at that famous Florida track in 1967.

what he saw, Fioravanti recounts, "but he said, 'Wider track, please.' I don't remember if it was 6 or 8 centimeters, but in any case it was enormous. More than in the tradition of Ferrari, where modifications were very small."

Another fundamental change made on the Daytona involved the use of a grille other than the traditional oval egg-crate type. While it wasn't easy to convince Enzo Ferrari to agree to the changes, perhaps the slow sales of the 275 GTB swayed him.

Fioravanti also designed the new GTB's interior, but, he explains, "Like the other Ferraris I did, there are no drawings of the interior. I put in the car what I got from the Ferrari people: the best driving position possible, the dials, gearbox, steering wheel, changing as little as possible." There were drawings for seat stitching and vent holes, but that was about it.

"Finally, we went for the official launch at the 1968 Paris Auto Show with a new metallic red," the designer continues, and the 365 GTB/4 ". . . was a very big success."

But the Daytona, its nickname derived from the 1-2-3 win of Ferrari 330 P4s at that track in January 1968, was also criticized. Its sin? A front-mounted engine. Ferrari already had the mid V-6 Dino 206/246—also with bodywork by Fioravanti—and the press expected a mid-big-engine Ferrari answer to the Miura, which was already two years old.

These harsh criticisms were expected at Pininfarina. In fact, many of the influential people at Pininfarina had been trying for years to convince Enzo Ferrari to do such a midengine car. "We discussed this with the Commendatore many times," Fioravanti recalls, "and we were ready to design a big Ferrari with a midengine, but the Commendatore was not absolutely sure."

What was his objection? "He was against it because the legend said the horse is in front of the chariot."

"This is going to sound dumb," Jay Leno begins, "but I think the most important part is its fatal flaw. The Miura is sort of like, well, you know there are guys who are white knights. They meet a beautiful waitress who has problems or a stripper who is screwed up and they just have to rescue them and save them. The deeper they get, the more in love they fall, the more problems there are. It's the same thing with the Miura."

Example?

"I've got an early production car, and every time you go around left-hand corners, the oil pressure drops, so you learn to how to deal with the problem. Consequently, it makes you more involved with the car.

"It's what I call the Betty Crocker theory of automotive involvement. In the '50s, the Betty Crocker people came out with a cake mix where all women had to do was add water and mix it up. They couldn't sell any of them. Then someone said, 'Why don't you make them break two eggs, add milk and water.' They did, and women bought them and were making cakes left and right because they felt like they were really baking a cake.

"When you're driving a Miura and it breaks down and you fix it and you get home, you feel like you've won a race. You're involved, and there's a great deal of satisfaction with that sort of thing.

"I have two Miuras, one being the early car, a 1967. It was Dean Martin's. He bought the car new for his son, who took it to school, went over a berm or something, cracked the crankcase, and the engine seized. Then it was given to a friend of mine who was a school teacher at the time. He couldn't afford to fix it. And, to show you how little Miuras were worth—this was back in the early '80s—he said, 'Do you want it? The engine's gone.' I took it, fixed the motor, and got

it running, but it's really a primitive car. The chassis flexes and things you wouldn't tolerate now, like no crash protection, aren't included. It's all done for the sake of good looks.

"Get going real fast—anything over 125 miles per hour—and you can actually sort of move the steering wheel on a straight road and the car doesn't turn. It gets that light and a little bit scary. Modern supercars have such high limits they aren't scary until you've crashed. You're driving along and you're going, 'Oh, I'm doing 180 . . . it doesn't seem like 180.' Whereas in the Miura you're doing 110 and, whoa, things are moving.

"The Miura is a much better car."

In its era, the Miura was a styling sensation, Leno remembers. "When you open an old *Road & Track* and see a Miura going down the highway in 1966 with the cars of the period . . .'55 Chevys, '57 Cadillacs . . . it looks like it could go under them. I remember seeing a picture in one of the magazines of a Miura on Sunset Boulevard at a light and all these enormous cars—even cars from the late '40s were around in 1965—and thinking, geez, what is that?

"And it cost $22,000, which barely gets you a Solstice now.

"I remember I was driving my Miura one day, and I looked in the rearview mirror and I go, 'It's raining . . . damn.' Then I looked out the front and it's not raining in the front. I realized what had happened was that one of the carburetor hoses had popped off and was spraying the rear window with gas. I pulled over, opened the back, and heard 'ping . . . ping' as gas hit the exhaust manifold. All I've got is this stupid little Haylon fire extinguisher. Luckily, the car didn't start on fire.

"People just don't have those sorts of adventures anymore."

While the design of the Daytona has become a classic, at the time of its debut, the critics were disappointed Ferrari hadn't done a midengine car like Lamborghini had. Leonardo Fioravanti was just twenty-seven years old when he took on the assignment to design the Ferrari that would become the Daytona.

Leonardo Fioravanti designed both the interior and the exterior of the Daytona.

Ferrari built fifteen competition Daytonas, while privateers produced others. While not designated as race cars, Daytonas proved solid runners in the GT class. Here, the Luigi Chinetti North American Racing Team competes in the 1973 24 Hours of Le Mans.

When the P400 (Miura) chassis was presented at Turin in late 1965, "we [Pininfarina] were completely destroyed. This was a chassis for a very innovative design," Fioravanti says. "We spoke immediately with the Commendatore. At the beginning he was even more contrary about a midengine car because Lamborghini was an ex-Ferraristi."

In March 1966, the complete Miura was unveiled. "I was very impressed," Fioravanti says. "This was a new car, like the Dino, but bigger." And yet Ferrari's resistance was such that it wasn't until 1968 that Pininfarina responded to the Miura with the Fioravanti-designed P6 midengine V-12 show car, which would prove to be the prototype for the Berlinetta Boxer.

Fioravanti has often ranked the Daytona as his favorite design, but he makes a somewhat typical designer's comment about this beautiful Ferrari. "When I finished this car in 1965, I was terrified. Horrible. Now when I see a Daytona I like it, but I see only the faults," he says.

In the Daytona design, Giorgetto Giugiaro found what he calls the huge legacy of a great sports edition of the period ". . . and it was a very rich period. It was an appealing and impressive sports car with a great emotional impact. It was really a reference to building a sports car, as was the Jaguar E-Type."

Road & Track (*R&T*) found both the Miura and the Daytona to be benchmark exotic cars. In its May 1968 road test, the magazine called the Lamborghini ". . . the most glamorous, exciting, and prestigious sports car in the world." Okay, the midengine Miura wasn't perfect, particularly the gearbox, but with 0–60 mile-per-hour time coming up on 6.3 seconds (the S version would cut that to 5.5) and a top speed of 163 miles per hour, the magazine concluded the midengine Lamborghini, at $21,000, ". . . has its faults but every enthusiast should have at least one Miura."

When *R&T* tested Ferrari's GTB/4, it admitted this car also wasn't faultless, but nonetheless concluded it was ". . . the best sports car in the world. Or the best GT. Take your choice; it's both." Maybe it was the $19,500 car's stunning looks, the 5.9 seconds in the 0–60, or the 173-mile-per-hour top speed; whatever the reason, *R&T* loved the car and took a shot at those who objected to its front-engine layout by subheading its Daytona road test: "The fastest—and best—GT is not necessarily the most exotic."

Let's update the Miura-Daytona controversy by dropping it in the lap of two famous American designers.

Freeman Thomas designed the Audi TT, moved to Chrysler to head its advanced design department, and is now the strategic design director for Ford Advanced Design. And he is a confirmed Porsche fan. He loves both cars, calling the Daytona "the last great hurrah of the great V-12 cars. The body design expressed that, though very modernistically. The elements are classical, the long nose and the cockpit set far back, but the graphics on that form were brand-new."

As for the Miura, it was "an argument against Ferrari. It was all about being rebellious. The styling of the car is nothing short of glorious," Thomas says.

The designer of Porsche's Boxster and the exterior of its Carrera GT is Milwaukee-born Grant Larson, who figures, "The Miura set a standard for future generations of midengine sports cars as far as the way the lines flow, especially the cutoff ducktail. It established a new form of language for midengine sports cars."

He admits to being a bit put off by the Daytona at first because "it represented a departure from my favorite era of Ferraris, which are those with more flowing lines up to about 1966. Regardless, I love the car because it has a very strong stance and it was a complete departure."

Which would you choose?

Maserati Bora 1971

163^{mph}

A driver can't complain that there aren't enough dials, switches, and levers on the dash of the Bora to keep his or her attention. The seats lay the driver out like a chaise lounge would but provide a very comfortable ride.

Though arguably not the sort of car that tickles the imagination like Ferrari's Boxer or Lamborghini's Countach, the Maserati Bora is an undeniably handsome machine. The nose suffered slightly from the US-mandated buck tooth "safety" bumpers that jutted ahead. Maserati's venerable V-8 resided behind the passenger compartment. The big carbureted twin-cam 4.7-liter engine, which had seen service in earlier race cars, was rated at 300 horsepower and had 325 lb-ft of torque in the Bora.

Thanks to its V-8, Maserati's Bora had a vaguely American sound that seemed to get more powerful the closer you got to the redline. This was 1973, and supercar acceleration wasn't as explosive as it is today, so you had to work harder when you drove the big Maser into a corner and then tried to top 100 miles per hour on the next straight. Work your way through the gearbox—can we make it to 100? . . . Now hard on the brakes.

The Bora was a 163-mile-per-hour car, but it took a while to get there, so we tended to play in the 70–110-mile-per-hour range, which was quick enough in the canyons north of Los Angeles with their gravel-strewn corners. When those cars slipped on the fallen scree, there was no intervention from electronics tapping brakes or pulling back throttle to save us; they just kept slipping. We wore seatbelts, but we didn't have an anti-lock braking system (ABS) or traction, and airbags were just a rumor.

Were we nuts?

Actually, Chuck Queener and I were in our late twenties, a couple of young magazine car guys thundering through the canyons in someone else's expensive (for then) Italian exotic car.

What, me worry?

It helped that the Bora had the unassuming torque of its V-8, so you could work your way up a tight and twisting hill road in just a gear or two, not having to spend much time shifting about to stay in a power band, as was necessary with some exotic cars in the early 1970s. Driving these cars quickly took a lot of work and concentration.

And the Bora did it without growling at you. Where other midengine supercars would let you know their powerplant was just aft of your head—sometimes to your pleasure, sometimes not—Maserati insulated the Bora's cabin more from the V-8. The engine sound was still there, but it accompanied you and never intruded. Like a respectful friend.

A very nice automobile, but . . .

It so happened that 1971 was the year of the Italian midengine exotic cars and, at the Geneva Auto Show in March, Lamborghini stunned the audience with the Countach, leaving pundits to speculate about how soon Ferrari would launch its big midengine model.

Over on the Maserati stand, almost lost in the speculation, the Bora was being tossed into the midengine mix, and if the Lamborghini and (expected) Ferrari were creating a splash, the Bora was kicking up ripples.

Too bad, because Maserati's early-1970s attempt at midengine stardom deserved better.

It certainly had the heritage.

The four Maserati brothers built their first race car in 1926 and produced competitive racing machines through the 1930s. When the Germans took over much of the European Grand Prix scene late that decade, Maserati found a home for its SCTF at the Indianapolis Motor Speedway, where

Wilbur Shaw drove one to win the 500 in 1939 and 1940. Masers were still competitive at the Brickyard after World War II, so when Ferrari was still just a rumor in the States, the name Maserati was familiar.

For all the corporate wealth seen on the exotic car front these days, in the decades after the war, things were economically tenuous for the likes of Maserati and Ferrari. In fact, the Maserati brothers had sold out to the Orsi family in the mid-1930s and stayed on to design and build cars, but no one was getting rich.

After the war, Maserati was successful with its small A6GCS sports cars and, later, the booming V-8-powered 450S. Juan Manuel Fangio won the Formula 1 championship in a Maserati 250F in 1957, and the company's innovative Birdcage race cars caused Ferrari fits . . . when they stayed together. It seemed that racing was the Italian specialist automaker's purpose for being, and road cars almost a necessary evil.

When the Orsi family wanted out in 1968, it sold its interest in Maserati to the French automaker Citroën, creating—at best—an odd couple.

Not wanting to miss the midengine exotic car craze, Citroën put Maserati's chief engineer, the highly regarded Giulio Altieri, to work on the project that resulted in the Bora.

Giorgetto Giugiaro was commissioned to do the exterior design, and his firm, Italdesign, summed up in its press release that Maserati wanted a car ". . . devoid of the exotic look that unnecessary decorations can create, strikingly sporty but not inordinately aggressive . . . innovative but not revolutionary." In other words, the French didn't want a Countach. That sort of flamboyance wouldn't be the Gallic way. By the way, how many great French exotic cars have you found in this book? And is the Bugatti really French or German?

Citroën certainly got its wish. No adult male's jaw dropped open, nor did any kid's knees go weak with delight upon seeing a Bora. Unlike the Lambo or Ferrari Boxer, teenage boys never decorated their rooms with posters of the big Maserati.

After the French automaker Citroën bought Maserati, it wanted to get in on the midengine craze. Citroën hired famous designer Giorgetto Giugiaro to design the body, but the car company made it obvious it wanted to avoid the glitz and glamor of cars such as the Lamborghini Countach. The Giorgetto Giugiaro–designed Ghibli, predecessor of the midengine Bora, was one of the prettiest sports cars of the 1960s.

Ferrari had the more famous name when the Bora was launched, but at one time, Maserati was better known to many racing fans in the United States. The Maserati 8CTF *Boyle Special*, driven by Wilbur Shaw, won the Indianapolis 500 in 1939 and 1940.

Nonetheless, it is a beautiful piece of sculpture. Conservative and a midengine next step from the front-engine Ghibli that Giugiaro also penned for Maserati, the Bora is possibly best known for its brushed stainless-steel roof and the fact that it is unerringly handsome. Whether that latter quality was enough for an exotic car was another matter altogether.

Maserati certainly stayed traditional with the engine, a 4.7-liter version of the venerable twin-cam sixteen-valve V-8 used way back in the 450S sports racing cars. In the Bora, it had 310 horsepower at 6,000 rpm and a sweet 325 lb-ft of torque at 4,200 rpm. For the Bora, Maser chose a ZF five-speed transmission with an upgraded shifter.

Maserati's V-8 was an impressive hunk of an engine under a large one-piece rear-hinged cover, a real mechanical beaut back in the pre-fuel-injection days when the V-8 was fed by four big Weber carburetors.

For the US market, Alfieri installed a 4.9-liter version of the V-8, which had been used in the Ghibli and was tamed for modernity and US emissions rules in 1973 to 300 horsepower at 6,000 rpm and 310 lb-ft of torque at 3,500. That nice torque and the sound of the V-8 gave the Bora an almost American feel and sound.

Maserati had to abandon its preference for De Dion rear suspensions with the midengine layout, opting for the traditional upper and lower A-arm design front and rear. The brakes were unusual in that, while they were vented discs, the 9.4-inch-diameter fronts were smaller than the 9.8-inch rears.

Citroën was famous for its master hydraulic systems and applied them to the Bora, not just for the brakes but also to adjust the driver's seat and move the foot pedals 3.5 inches front or back. Though not so unusual these days, this ability to move the pedals, combined with the steering column's fore/aft and up/down adjustability, made the Bora amenable to just about any driver's physique.

You would expect to see the Bora's seats in a *Jetsons* episode—lovely long shells with leather upholstery you could settle into and that would support you well out to under your knees. The dashboard had an array of eight gauges—with a big tachometer and speedometer—and a row of rocker switches that placed the Bora in the early 1970s.

Road & Track tested a 4.7-liter European Bora that had been certified to US standards at an estimated 300 horsepower and 325 lb-ft of torque. It wasn't a light car, with a curb weight of 3,750 pounds, and it was geared for a top speed of 163 miles per hour; nowadays its 0–60 mile-per-hour acceleration of 7.2 seconds doesn't look all that impressive.

When it went on the market in late 1971, the Bora's price was around $25,000, but by the time it was certified for the United States with the dollar dancing around European currencies, the price elevated into the mid-$40,000 area. Apparently, between 1971 and 1979, Maserati produced 530 Boras.

While Ferrari had its big twelve-cylinder Boxer and its distinctly different, smaller V-6 and then V-8 models, the Bora had a more direct little brother. Called the Merak, it shared the chassis and much of the body of the Bora, with its own flying buttress upper rear bodywork. Little brother inherited much of the equipment of the odd-but-interesting Citroën SM, including the instrument panel, the strange little bulb brake pedal, and the SM's Maserati V-6 front drivetrain turned front-to-back and mounted aft.

Priced at about $10,000 less than the Bora, the Merak wasn't very quick—about 9 seconds to 60—but it was a comfortable machine.

Boras weren't necessary automobiles to fall in love with, but they were cars to be appreciated. Unfortunately, exotic cars need more than that for most drivers. They need to get your heart beating faster, give you a little rush, leave you a little breathless.

Maserati's Merak, the Bora's little brother, borrowed much from the Citroën SM. With a mid-mounted V-6 engine, the Merak cost about $10,000 less than the Bora.

Ferrari finally made the move to a midengine sports car layout when it created the
365 GT4/BB, the last two initials standing for Berlinetta Boxer. The prototype was
presented at the 1971 Turin Auto Show, while the car went into production in 1973.

Ferrari Boxer 1973

188mph

Enzo Ferrari was under fire. Lamborghini had set the exotic-car world aflame with the midengine Miura in 1966, and here it was, mid-1971, and there still was no twelve-cylinder midengine road-going Ferrari.

British race car constructor John Cooper had been the main force in the modern European move to midengines in the 1950s and had done the same at the Indianapolis 500 in 1961. Throughout the 1960s, sports racing cars had shifted the engine to the rear, and the trend was being taken up by high-performance sports cars for the streets . . . but not yet by Ferrari.

Why go midengine? As winning race cars were proving, there were significant handling advantages and, with the engine in the back, the frontal area could be cut significantly. It made more sense from an aerodynamic standpoint, and it was easier to fit big rear tires on midengine machines. There were disadvantages, such as minimal luggage space and poor visibility to the rear, but they were not crushing problems in exotic cars used only occasionally. Just as important, perhaps, was image. Midengine was the proof you were technically advanced; it was the next new thing.

Enzo Ferrari would have none of it in his road cars. He had always been conservative about engineering changes and was reluctant to adopt midengine technology. He'd resisted the move to disc brakes in the late 1950s, and Ferrari was the last major Grand Prix constructor to situate the engines of its race machines behind the driver. There had also been a slow transition to midengines with his sports racing cars, and it seemed he was making his point in June 1962 when Phil Hill and Olivier Gendebien won the 24 Hours of Le Mans in a front-engine Testarossa.

Ferrari always made the necessary changes eventually, and when he did, he won. Hill drove Maranello's initial midengine GP car to a World Driver's Championship in 1961. And the midengine sports cars were quickly competitive, from the 1961 246 SPs, through the beautiful 330 P4s in 1967, to the victorious 312 PBs in 1971–1972.

Ferrari built the Pininfarina-designed Boxer at the Scaglietti factory in Modena, Italy. At the time, Scaglietti (now owned by Ferrari) was an independent constructor of cars for Ferrari and other car companies.

It isn't just great automobiles that separate Ferrari from others, but the fact it is the only automaker that fields its own winning Grand Prix cars, building both the engine and the chassis. In 1983, the Boxer's GP companion was the 126 C3 with a carbon-fiber/Kevlar frame and a turbocharged 600-horsepower 1.5-liter V-6.

But where, pray tell, was the twelve-cylinder, midengine Ferrari road car?

It was at the Turin Motor Show in the autumn of 1971 as a concept car. This was another battle in the turf war being fought between local rivals Ferrari and Lamborghini. One-time Ferrari customer Ferruccio Lamborghini had again stunned the automotive world the previous spring by showing Bertone's dramatic Marcello Gandini–designed Countach LP500 show car at the Geneva Auto Show. Ferrari shot back that fall in Turin with the Berlinetta Boxer, shaped at Pininfarina by Leonardo Fioravanti on the theme of the 1968 P6 show car.

As it was, the 365 GT4/BB (BB for Berlinetta Boxer, though its common name is simply Boxer) needed two more years before deliveries began in 1973, while the Countach wasn't ready for customers until 1974. But enthusiasts certainly had something to talk about in the meantime, and both final production models were almost identical to the show cars.

From a styling standpoint, the Boxer was an interesting contrast to the Countach. There is no denying the Lamborghini is spectacular, jaw-dropping, cool, and memorable, looking as though it could go nose-up and head for the moon. Arguably, however, the Boxer is more beautiful and ultimately more sensual.

Its designer was Leonardo Fioravanti, and if you worship Ferraris of the late 1960s and early 1970s, he is your idol. He penned the Daytona in his twenties and went on to shape the Dino, Boxer, and 308 GTB. And he's still at it in his sixties, working independently and getting credit for the clever folding glass top on the 2005 Ferrari 612 Superamerica. Fioravanti also designed Pininfarina's 1968 P6 show car and made the production version of it into the Boxer.

With a long nose, great haunches, and a flying buttress rear window, the Boxer had a more graceful form than the aggressive Countach. It was also rather easier to drive, as there wasn't so much automobile around you, so you didn't feel quite as encapsulated. It also had rather large windows that made outward visibility better than in other exotics.

The interior was just what you'd expect in an exotic car: easily read gauges set deep enough to avoid glare, and switchgear that, today, looks a bit archaic, but took up little room and, for the most part, had a firm, positive feel. As an example of how simple things were then, the sun visors rolled up and down like window shades and, when unrolled, were held in place by suction cups. In fairness, road testers of the day complained about the interior's lack of flexibility; it forced the driver to adapt rather than the other way around. No eight-way power seats or adjustable lumbar supports in the Boxer.

It was important that Ferrari's mightiest road car, the Boxer, should have a twelve-cylinder engine. There was a mid-mounted V-12 in the P6, but the BB had a flat-twelve, otherwise known as a "boxer" engine; hence the car's name. Ferrari race cars had been using flat-twelves since 1964, and they were particularly successful in the early 1970s with the 312 PB race cars, which won the World Sports Car Championship in 1972.

Why a flat engine? Its shortness lowered the aerodynamic profile of the car and put the engine weight lower in the chassis, dropping the center of gravity.

Although Ferrari adopted the flat-engine layout for the Boxer, it didn't adapt a racing engine to the production car. Instead, engineers basically took the company's aluminum V-12, with its traditional 60-degree vee between cylinder banks, and opened that angle to 180 degrees.

As the car's official designation, 365 GT4/BB, suggests, the Boxer had the same 4.4-liter displacement as its predecessor, the 365 GTB/4 Daytona, even keeping the 81x71-millimeter bore and stroke and 8.8:1 compression ratio of the V-12.

It was an impressive engine, particularly with the air cleaners off, its two banks of paired three-throat Weber carburetors standing rather proudly atop it. Where the factory claimed 352 horsepower for the Daytona's V-12, it pegged the Boxer's at 380.

Just as the Countach had a unique engine-transmission layout with its five-speed ahead of the V-12, the Boxer was also unusual in that the five-speed was located below the flat-twelve; it was a part of the engine.

The frame of the Boxer was a cage of steel tubing, and while the Boxer's hood, rear deck, and doors were aluminum, the remainder of the body was steel. Its suspension had upper and lower A-arms at each corner with coil springs and tube shocks. The brakes were discs inside Cromodora wheels.

As exciting as the Boxer might have been, it wasn't meant for the United States, in which Ferrari never officially sold the car. In the early 1970s, huge automakers, such as General Motors and Ford, were struggling with new laws meant to lower engine emissions and increase road safety, but small companies such as Ferrari and Lamborghini were downright flummoxed by the rules.

Other men weren't. As automobiles such as the Boxer and early Countach were kept from the United States, a small industry was created to certify

Inside, the Boxer was typical for its time, with complete instrumentation set tightly in a pod and the expected tall shift lever.

To power the Berlinetta Boxer, Ferrari engineers took the company's famous 60-degree V-12 and opened the angle between cylinder banks to 180 degrees, making it a flat or "boxer" engine, a layout the factory had been using on race cars since the 1964 512 Formula 1 engine.

This is one of the factory-built silhouette 512 Berlinetta Boxer race cars. Compared to the street 512 BB, the LM (Le Mans) versions had Pininfarina-smoothed bodywork, added inlets, a rear spoiler, and an extra 16 inches of length. This car was the first sold to an American, Bob Donner; it finished ninth overall at Le Mans in 1981. In addition to being lighter than stock at somewhere shy of 2,400 pounds, the racing Boxers got a horsepower increase to 480–500 horsepower, which was enough to propel the spectacular-looking Ferrari to just over 200 miles per hour.

them for sale there. It was complicated work. New structures had to be created inside bumpers to absorb the shock of mandated crash tests, hopefully without making the cars' ends butt ugly. Huge Weber carburetors had to be returned, camshafts often reground, and catalytic converters plugged into exhaust systems to get engine emissions down to the federal limits.

Some of the men who did these conversions were very clever, and their work was impeccable. Some were scam artists. Some managed to make the conversions without adding huge masses of weight to the cars' ends (not the best thing for handling) and draining too much horsepower. Some made a mess of the safety bumpers and created monsters that met the legal emissions standards but would barely run in everyday traffic.

The feds kept an ever more vigilant eye on this business of "gray market" cars and, by the mid-1980s, they had almost eliminated it with ever stiffer laws and penalties. By the time the very desirable Porsche 959 came along in 1989, no one was even trying to certify it. That would wait for new rules at the turn of the century—more on that in the chapter about the 959.

Back to the Boxer. In 1976, Ferrari increased the displacement of the flat-twelve, opening the bore 1 millimeter and the stroke 7 millimeters to

create the 5.0-liter Boxer 512. It also changed its number system: while the 365 designation in the previous model referred to the displacement of one cylinder (12x365 = 4.4 liters), the new one meant a 5-liter, twelve-cylinder engine.

Anti-emissions laws were taking hold throughout Europe too, lowering pollution and horsepower. So, despite the greater displacement and 9.2:1 compression ratio, the 512 BB had 360 horsepower—down 20 from the 365—but with about a 10 percent increase in torque. Ferrari also made this version a dry sump engine to lower it a bit in the car to take better advantage of the flat engine's low center of gravity. The car was also a bit wider at the back to accommodate rear tires that were now larger than the front ones (215/70BR-15 front, 225/70VR-15 rear). There were also a few exterior changes, such as a new front spoiler and added ducts.

Road & Track tested both 4.4- and 5.0-liter Boxers in the United States, the former uncertified (and, as a result, getting the owner in trouble with the feds when they saw the road test), and the latter nicely reworked to meet the rules. Both cost around $40,000 new, while the 512 BB had some $45,000 in conversions to make it legal.

For the sake of the 365's clutch (and owner), the tester had to ease the car off the line, so the 0–60 time of 7.2 seconds was neither impressive nor truly representative . . . but that was the way supercars were in the early 1970s. They didn't like drag-racing starts, being much happier ripping their way past 100 miles per hour. These cars were happiest when run at or near maximum velocity.

The 512 was up to the task and got to 60 in 5.5 seconds. The 365's top speed is estimated at 175, the 512 at 188, and I know for a fact that the latter achieved 155 with little strain and with plenty left. The guy who did it for *Road & Track* on a quiet Orange County highway just couldn't wipe the smile off his face.

R&T did another test—one of its most famous—when it pitted a 512 BB against Countach at the high-banked Transportation Research Center

in Ohio. Both cars were legalized machines, so they were a bit down on factory-claimed power but proved quick nonetheless. *R&T* cut the Ferrari's 0–60 time down to 5.1 seconds against the Lamborghini's 5.7 seconds, after which race driver Sam Posey took the Boxer to 168 miles per hour and the Countach to 150.

With another nod at still stricter emissions laws around the world, Ferrari did one more variation on the Boxer: the 1981 512 BBi, the little "i" meaning it now had Bosch K-Jetronic fuel injection. While this helped meet the rules and made the cars easier to maintain, it cut another 20 horsepower, bringing the 5.0-liter engine down to 340.

This would be only a three-year model, leading up to the Testarossa.

In a Boxer, it always felt odd that the tall shift lever and the slim rim (by today's standards) steering wheels should be attached to such powerful stuff. It might have been the fact that this engine seemed to growl more than most exotic cars, sounding as though it was trying to attack you from right behind the cockpit.

Today's supercars are meant to feel like they are of one chunk—hefty, buffed up, and packed like a fighter jet with electronics that will help you go faster, yet save your backside if you get stupid.

The Pininfarina midengine design set the pattern for a generation of Ferrari sports cars. Pininfarina has been the designer of almost all major Ferraris since the mid-1950s. Aerodynamic development in Pininfarina's wind tunnel was integral to the design of the car, and was one of the first modern examples in Europe.

The Boxer had a more athletic feel to it, less like a power lifter than a sprinter. You didn't so much power off the line as let the engine wind out while the revs climbed. Exotic cars then weren't as blindingly quick or sudden as they are today—not so much immediate gratification off the line. Instead, they tested you down the road as it bent left and right, rose and fell, the car on narrow, less grippy tires and with no electronic aids to save your bacon if you screwed up. You didn't go as fast in those days, but to be truly quick, you had to be more talented than today.

Sorry, boys, but that's the truth.

Even though the Boxer wasn't sold in the United States, the first attempts to race a 365 FT4/BB were based there. Longtime US Ferrari importer and three-time winner of Le Mans, Luigi Chinetti prepped a pair in the mid-1970s, but other than a sixth at Daytona in 1975, the effort wasn't terribly successful.

The factory stepped up in 1978 to build three 512 BB LMs (Le Mans) for that French race. With only minor body modifications, they were good-looking race cars that had shed weight down to around 2,700 pounds and had 440 brake horsepower, though none finished the race.

Pininfarina got involved the following year when regulations allowed GT race cars more variation from the stock versions. Shaped in Pininfarina's wind tunnel, these Series II and III 512 BB LMs were wider and 16 inches longer—mostly at the tail—and had smoother, much more aerodynamic bodywork. Horsepower was upped to 480–500, weight was cut still more, but as spectacular as they looked and despite a top speed of just over 200 miles per hour, they were never as successful as they looked like they should be.

In 1984, Ferrari replaced the Boxer with the Testarossa, retaining the mid-flat-twelve layout. Still a supercar, it was fully certified for sale throughout the world. Many of us went "Wow" when we saw the Testarossa's Pininfarina styling, which was spectacular and somehow seemed to cater to the inner tingles that made Countachs so popular.

A great car, but somehow it made us miss the Boxer.

This is the second Countach prototype and the oldest existing Countach. It was originally red, but Lamborghini repainted the car green for presentation at the 1973 Paris Auto Show.

Lamborghini Countach 1974

180^{mph}

Today there's too much traffic to rush along in front of the Lamborghini factory, but not in 1988. Then there was plenty of open space for Sandro Munari to show me how Lamborghini test drivers scared the daylights out of the clients of their 180-mile-per-hour Countachs for years.

Munari is a four-time world rally champion and one of the world's great car control experts. For several years, he also led Lamborghini public relations, which is why he was driving the Countach as we hurried down the dead-straight road toward Modena.

We left the factory briskly, to say the least—down somewhere around the 5.0-second 0–60 time—and kept up a good head of steam as we crested a small rise in the road. The Lambo felt a bit lighter and then . . . *good grief! There's a hard left turn ahead!*

I knew it was coming and Sandro could handle it, but I gulped nonetheless. Hard on the brakes, he drove us forward into our seatbelts and pitched the big Countach left through the tight turn, smoothly caught it and, bang, back on the throttle. The hunk-of-a-V-12 was wide awake and we squirted down a short straight to a 90-degree right where Sandro deftly nailed the apex, and we were quickly out the other side, flying toward Modena.

I had always wanted to do that.

In automotive history, there have been exotic cars with more impressive engineering, a flashier time to 60 miles per hour, a greater technical portfolio, and even higher top speeds than the Lamborghini Countach, but none have captured the public's attention like the big Lambo.

Stunningly sleek as a show car, tarted up in its later years, the big Lamborghini stole and broke hearts around the world. It was the poster boy for all supercars; it was their Elvis.

With the Miura, Ferruccio Lamborghini had begun seriously sparring with Ferrari in 1966, and now he came back with what might have been a serious body blow if only it hadn't taken so long to land the punch.

Lamborghini had outmaneuvered Ferrari by showing the midengine Countach LP500 at the 1971 Geneva Show, six months before the similarly laid-out Berlinetta Boxer debuted. Even before the public saw the big Lambo, it had apparently stunned onlookers. Other Lamborghinis had been named after famous fighting bulls. *Countach* is a somewhat off-color Piemontese (language spoken by many people in northwest Italy) expletive of surprise uttered by one of those who saw the car early on.

Marcello Gandini takes the credit for that surprise, having shaped the Countach while working for the design firm Bertone. This wasn't a one-trick pony for Gandini, who also gets credit for such important automobiles as the Maserati Khamsin, Fiat X1/9, Ferrari 308GT/4, Lancia Stratos, and the car that first used (now famous) swing-up doors, the Alfa Romeo Carabo.

This taillight is the view most drivers saw of a Countach in its early days. Few automobiles have managed to match the public impact of the Countach.

Like any true big Italian exotic car of its day, the Countach has a V-12, which began as a 4.0-liter. It then was increased to 4.7 liters and then to 5.2. The latter version had four valves per cylinder, and horsepower rose to 455. The engine sits "backward" in the car, its transmission jutting forward into the cockpit.

Those first LP 400 Countachs earned their glory—so low and wide, arcing strongly from front to back, the two-seat jet fighter cockpit, just enough louvers, Gandini's signature forward-leaning rear wheel arches. Later, Lamborghini would yield to faddish spoilers and other add-ons that corrupted what Gandini first drew.

BMW's design director Chris Bangle echoes the feelings of many of us when he says, "A Countach in anything other than its original form is so *Pimp My Ride* it's almost impossible to discern the original."

Bangle refers to supercars like the Countach as having a look of "post-atomic technology. "He asks, how did Gandini do cars like the Lamborghini Countach or Bertone Stratos show car I thought were so cool?." And then he explains: "He took the things you could exaggerate—like the wheel-to-car relationship, the lowness and wideness of the car—and exaggerated the hell out of them. Getting in and out of the Stratos [concept car]—a car that could drive underneath my chair—would be difficult, but you were expected to go through some agony to get in and out of them, and that agony was pleasurable."

To Bangle, true supercars are those that shake us—that make us take extra notice. "Cars that are so politically incorrect, [with] no possible reason for being, they are the supercars," he says.

Gandini drew his inspiration for the big midengine automobile from the sports racing cars of the day. He thought that race cars had changed in the late 1960s, and while they had improved, they had become more functional and were no longer beautiful for the sake of beauty. They needed lovely lines less, aerodynamics more—to keep them from flying off the road at high speed. Such race cars became wild-looking, in the manner of the long-tail Porsche 917s. Gandini said, "If there was a car that inspired the Countach a little bit, it was the Lola T70 Coupe." With the new Lamborghini, the designer went for a feeling of these sports racing cars and how they pushed their mechanical parts to the limit. Gandini wanted "people to be astonished when they see the car."

He certainly succeeded.

Getting into a Countach means stepping over a wide sill and then dropping down into a reclining leather-upholstered seat. The shift lever is attached directly to the forward-facing transmission.

This 1976 Countach is the super Lamborghini in its cleanest form, before extra spoilers and wings sprouted. BMW design chief Chris Bangle says Countach designer Marcello Gandini " . . . took things you could exaggerate—such as the wheel-to-car relationship, the lowness and wideness of the car—and exaggerated the hell out of them. Cars that are so politically incorrect, with no possible reason for being . . . they are the supercars."

Styling wasn't the only unique feature of the Countach. Chief engineer Paolo Stanzini (later of the Bugatti EB110) had this big V-12 to fit in the chassis. The Miura had a sideways transverse V-12, which was great for a short wheelbase but could be a wide package, with one cylinder banked up close to the bulkhead and a need for a very specific, very expensive drivetrain to get the power to the rear wheels.

A north-south longitudinal layout, with the engine ahead of the gearbox, is common in midengine automobiles but lengthens the wheelhouse and puts accessory drive belts next to the firewall.

So Stanzini turned the engine around, with the accessory belts to the back and the transmission sticking forward into the cockpit's center tunnel, with the clutch between the V-12 and the five-speed. This layout had several advantages, such as shortening the wheelbase and putting the shift linkage right on the gearbox.

One problem: how did he get the power from the five-speed gearbox rearward to the differential? Stanzani routed it off the transmission countershaft to a shaft that spun in bearings under the clutch and transmission, then through the engine's sump to the limited-slip differential, which was in its own compartment aft of the engine's sump.

Over the Countach's 1974–1990 lifespan, the V-12 was in three power forms. Its original 4.0-liter edition had 375 horsepower at 8,000 rpm and 387 lb-ft of torque at 5,000. When displacement increased to 4.7 liters, the horsepower stayed the same (remember, these were the power-killing emissions years), but torque rose a bit to 409 at 4,500. The increase to 5.2 liters also then brought four valves per cylinder and horsepower progressed to 455 at 7,000, torque to 502 lb-ft at 5,700 rpm.

It all depended on gearing, of course, but 0–60 times started in the mid-five-second range and were down near the mid-four-second range with the four-valve engine. Top velocity varied from the 175-mile-per-hour area up to almost 200 miles per hour, with the winged versions generally more stable but slower.

Under the Countach's amazing aluminum body was a tubular space frame. Under that was the suspension, upper and lower A-arms with coil springs and tube shocks at the front, while in back were upper links, a reversed lower A-arm, and a pair of coil springs and shocks per side. Vented disc brakes were used all around, while the tires—at least on the early cars—looked narrow for a car with so much potential.

Those flip-up doors were more than just a design gimmick, because conventional swing-outs would have been impossibly long, and that was true even if you weren't in a parking lot. Even with them, it was a stretch to get in and over the wide sill, so you tended to climb over and drop into the seat, which laid you out like a chaise lounge. Getting out was more difficult.

What you faced was a short but wide rectangular instrument panel and seven purposeful gauges. Radio and climate controls were another stretch forward at the front of the center console. There was no mistaking the car's huge size as you stretched again to reach up and pull down the door.

 The earliest versions of Lamborghini's Countach could hit 60 miles per hour in the mid-five-second range, while the later cars did the same in the mid-fours. Top speed for the midengine supercars ranged from 175 to almost 200 miles per hour, depending on engine and body configuration.

It's a good thing the Countach was fast so you could always stay ahead of most automobiles, because rearward visibility was minimal. I always envied factory mechanics who had a marvelous trick for sitting on the driver's sill, door up, and looking backward as they worked the pedals and steering to reverse a Countach. I have never been that dexterous.

There was big stuff rumbling behind you at the turn of the key, and when you moved the shift lever down and left into first, it was like an extraordinarily hot knife cutting through an anvil. You were always aware that it was a large, wide car, weighing as much as a National Association

"The sexiest cars have a male and female side to them," Jay Leno explains. "The Miura's engine is very masculine, but its styling is very sensual. The Miura and Jaguar XK-Es and XK120s are cars women think are very sensual. As for the Countach, women may go "oh" or "ah" when they see one, but, Leno says, "I've never seen a woman think a Countach is attractive.

"If you're trying to impress twelve-year-old boys, the Countach is the greatest car in the world, because they jump up and down and go crazy. I would get letters from kids writing that they knew I had a Countach and said, 'If you could take me to school one day, it would be the coolest thing.'

"So I wrote back and did that a couple of times. One kid lived in Huntington Beach. I picked him up in the morning, pulled in front of his school, the door goes up, and all his friends are watching. I said, 'So long, Jimmy.' It's fun doing that, and in that sense the car is great.

"The Countach suffered from what a lot of the early supercars did, in that most of them were fairly primitive, with a tube chassis. It's a big, heavy car that's not nearly as fast as it's supposed to be. I guess in its time it was pretty quick. Now it seems fairly agricultural. It is fun to drive and I enjoy it. It's just not modern fast."

Originally, Leno's Countach had a rear wing, but, "I didn't like the wing, so I took it off. I mean, the chances of you getting airborne on the street are fairly remote.

"I like the Countach, but not as much as the Miura. I would have preferred to see the Miura perfected—the next generation of the SV."

for Stock Car Racing (NASCAR) stock car, but that didn't take away from the fun. Okay, it wasn't the most refined supercar in some respects, and you never forgot all the power behind you, but it went like the wind. Besides, you were driving a Countach, poster boy of the exotics, and that alone was worth a lot.

Countachs have been at the center of so many stories about aggressive driving. Here's my favorite:

Tom Bryant would become editor-in-chief of *Road & Track*, but when he was a young editor he was assigned to write a multi-exotic-car test, which included the Countach. The crew headed for the desert and a quiet, isolated public highway loop to do some high-speed comparison testing. Every lap, they sedately passed a highway patrol officer, who finally couldn't stand it and stopped the Countach and all the cars behind it.

Uh-oh. The highway patrol officer looked the Lamborghini over, walked over to Tom, and said, "I don't know what you're up to, but *stop it.*"

They did. Wouldn't you?

 As much as anything else, the Countach is famous for its flip-up doors, which are now adapted to custom cars and trucks. They aren't just a visual attraction; conventional doors on a car of the Lamborghini's size would have been quite long and difficult to open and close.

 As the years progressed, the Countach design was modified with added wings, spoilers, and lower sill extensions, as on this 1988 version. To some fans, the additions were cool, while to many others they tarted up the basic excitement of the Countach design.

Porsche has often made special editions of its Turbos, such as this 1993 lightweight model. Although the basic shape of the 911 Turbo from its first iteration in 1975 to the latest might look unchanged to an outsider, the exterior design has been through many variations.

Porsche Turbos 1975

155^{mph}

When the speedo needle slipped past 150 miles per hour, I knew everything was going to be okay. Up on the banking at what was then called Transportation Research Center, off in the Ohio countryside, the Porsche Turbo felt snug and secure at speed, maxing at 155.

I'd already had a chance to rush the Turbo to 60 miles per hour. It needed some 6.7 seconds, which isn't quick today, but there were extenuating circumstances. We were just learning how to spool up a big turbocharger off the line, and even when we learned how, it could be tough to launch a Turbo.

Who cared? This was 1975, and after a half decade of watching horsepower in performance cars disappear like smoke, thanks to (badly needed) anti-emissions laws, the Turbo gave me the feeling that there was hope after all. Speed was back.

It's a fact that when you mention going quickly in an automobile, the subject of Porsche Turbos is close at hand. No one is too picky about which vintage is mentioned. No big concerns about 930 or 996 or rear-wheel drive or twin turbos or the size of the tail. It's just that Porsche Turbo equals speed. Simple as that.

There are entire books devoted to the subject, so think of this as paying homage to the type—to more than thirty years of Turbos.

That first supercharged 911 was a two-year model, the 3.0-liter version of the famous flat-six equipped with a big turbo that boosted it to 234 horsepower at 5,500 rpm and 246 lb-ft of torque at 4,500. Even with that now-middling 0–60 time, it felt like a handful, and heaven help you if you thought you could get on or off the throttle quickly in a corner.

Who cared? We loved the Turbo, and it came with that great "whale tail" rear spoiler.

During 1978 and 1979, Porsche dramatically picked up the pace with a revised Turbo, which had a displacement increase to 3.3 liters and, possibly more important, an intercooler neatly designed into the whale tail. Horsepower got a nice boost—so to speak—to 253, torque to 252 lb-ft, but it was a much quicker car, to 60 miles per hour in just five seconds while we hung on for what felt like a rocket booster lighting off just shy of 4,000 rpm.

And then Porsche abandoned us. They pulled the Turbo from the United States, and it was even rumored the 911 series was doomed by emissions and such. How depressing.

Those of us in the States were supposed to be happy with 928s and 944s while Europeans still got Turbos, which were headed for 300 horsepower. Some Americans weren't put off and bought gray market Turbos and had them certified (or at least claimed to be) by private firms.

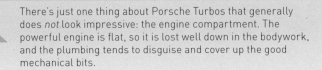
There's just one thing about Porsche Turbos that generally does *not* look impressive: the engine compartment. The powerful engine is flat, so it is lost well down in the bodywork, and the plumbing tends to disguise and cover up the good mechanical bits.

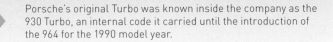
Porsche's original Turbo was known inside the company as the 930 Turbo, an internal code it carried until the introduction of the 964 for the 1990 model year.

Peter Gregg, driving for the Brumos team, was one of the finest Porsche drivers in US racing history. He won more than forty races in the 1970s, eight of them in 1979 driving this twin-turbo, 700-horsepower Porsche 935.

By 1984, the Turbo had a 3.3-liter engine, an intercooler, and enough power to get to 60 miles per hour in 4.9 seconds—but not in the United States, as emissions control problems kept the car out of that market from 1980 until 1986. Seen with this car are World Driving Champion Phil Hill (left) and Le Mans winner Paul Frere.

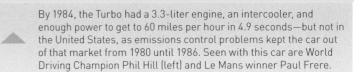

What says *Porsche Turbo* better than the big whale tail with its intercooler? This 1988 version had the 300-horsepower 3.3-liter engine.

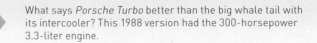

Porsche will do special modifications of its cars for customers. Requests have run from installing gold shift knobs to special leathers to more power. In the mid-1980s, it was common for owners to have the factory convert their Turbos to the flat-nose look.

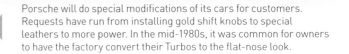

We were saved by an American, Peter Schultz, who ended up in charge of Porsche and ordered the Turbo back into the States, where it returned in 1986. The engine was still at 3.3 liters in the 930 platform, and while horsepower was up by almost 30, antismog laws had kept performance from improving. But we had the Turbo back and that counted for a great deal.

The 944 version of the 911 Porsche came along in 1991, still at 3.3 liters, but in 1993 it was increased to 3.6 and things got serious again. The 0–60 time was down in the mid-four-second range, and Porsches were being tamed, losing their reputation for getting tail happy.

Come 1995, Turbo horsepower was at 400 and Porsche put all-wheel drive under the car to make things even more manageable. The 0–60 times were under four seconds, and top speeds were in the 175-mile-per-hour range.

With the 996 design, the Turbo took another step forward, and while performance numbers weren't all that different, it had such technical advances as Porsche Stability Management; now Porsches were eminently manageable.

Ratchet forward to early 2006 and the 911 Turbo based on the 977 chassis. Turns out the engine is still at 3.6 liters, but look what they did: 480 horsepower at 6,000 rpm and 457 lb-ft of torque on a broad base . . . if you don't buy the Sport Chrono Package that extends boost for an added ten seconds and drop-kicks torque to 502 lb-ft.

Ach du lieber!

The all-wheel-drive system is now lighter, the electronics refined again, the brakes still larger, and a very tricky variable turbine geometry system changes the pitch of the turbo's blades to best take advantage of the exhaust flow.

So now we are looking at a 0–60 time of 3.6 seconds with a manual six-speed, a tick quicker with Tiptronic if you have time to notice on your way to 195 miles per hour.

That's a Porsche Turbo for you: a hero car for more than three decades. And we haven't even touched on the fact that Porsche put its racing where its mouth was; its Turbo was an assumed winner around the world for decades.

Several years ago, *Road & Track* was invited to Germany to drive 911s from over the years. Included in the stable was an original Turbo. We had an 8-mile circuit through the forests near Zuffenhausen and the chance to let the cars go. After two laps in the original Turbo, I asked Klaus Bischof, the amiable leader of Porsche's historic section, "Is that car really up to spec?"

I couldn't imagine Klaus giving us anything save a perfect car, but the Porsche felt so slow. He laughed and said I wasn't the first to ask that question . . . and, yes, the car was in excellent tune.

To a modern right foot that felt any sports car over 0–60 in five seconds was lagging, the vintage machine was just that. But to think of any Porsche Turbo as anything but a hero car would be as historically lame-brained as dismissing Babe Ruth, Bill Haley and the Comets, or Chuck Yeager as irrelevant.

By the mid-1990s, Porsche had an all-wheel-drive version of the 993 Turbo and 400 horsepower and 400 lb-ft of torque on tap. This "oomph" came from a 3.6-liter version of the famous flat-six with a pair of turbochargers and intercoolers.

Imagine having this red 1994 Turbo out in the Arizona desert with plenty of straight road and not a soul in sight. This version puts 360 horsepower under your right foot, thanks to an increase to 3.6 liters.

The interior of a 1988 911 displays classic Porsche design.

According to BMW's testing, a new M1 would get to 60 miles per hour in 5.3 seconds, and its top speed was a few ticks over 160 miles per hour. While not necessarily impressive today, those numbers got plenty of attention in 1978 when the BMW supercar was launched at that fall's Paris Auto Show.

BMW M1
1978

162^{mph}

There is a vintage, welcomed, and honest sound to the BMW M1's six under acceleration. Much of it seems to be induction noise, the sound of air being drawn into the injector rams, with some exhaust rap. For being so close behind the cockpit, the engine's sound isn't obnoxious, and two people can converse without raising their voices too much.

If you're driving fast, you'd probably be too busy to talk. Supercars in the late 1970s didn't have the ability to unleash today's mega horsepower. There was no point, push, and hang-on driving with electronics to save your bacon if you screwed up. In a BMW M1, it was a more about a balanced interplay between power, steering, brakes, tires, and handling, which had to be treated with more respect than today.

You built momentum and then maintained it. Nowadays, you can zip through a supercar's gearbox and be at 130 miles per hour with ease. In the BMW M1's day, you earned 130 miles per hour.

For all the supercars BMW has built, from snarling little 2002 tiis in the early 1970s to modern V-10 M5s, it has only done one automobile you could officially classify as a supercar: the midengine M1.

It was the mid-1970s, and BMW was already racing the 3.5-liter CSi coupes, which were wonderful to watch, always a bit out of shape as their drivers pushed them to the limit . . . and still not quite quick enough to stay with Porsche's latest.

BMW clearly wanted to beat its German archrival in the international Group 4 and 5 racing classes but needed a new weapon. And that new BMW would have to be homologated for competition, which meant building around four hundred examples of the car within months. Rather than trying to make a race car out of another of its production cars, BMW went the other way, creating a race car that would then be productionized to meet the homologation rules.

BMW management signed off on the program, but what followed was a classic case of best-laid plans going awry.

Already too busy to design and build the M1 on its own, BMW farmed out much of the project. While the all-important engine would stay with BMW Motorsports (who better?), the rest of the project was sent south to Italy.

Giorgetto Giugiaro at Italdesign was contracted to design the body. There had already been one sensational midengine BMW, the gullwing-door 1972 Turbo created by well-known designer Paul Bracq to celebrate the (now infamous) Summer Olympics in BMW's hometown, Munich. There were hints of Bracq's car in the M1, particularly in the rear three-quarter view, but the M1 had more in common with other cars being designed in that era by Giugiaro, such as the Lotus Esprit and the DeLorean.

Lamborghini was contracted to develop the chassis, which assigned design responsibilities to one of the most prolific of all chassis men,

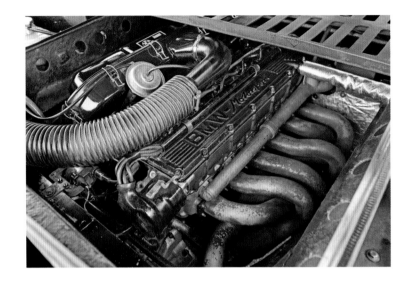

BMW Motorsports developed and built the twin-cam straight-six engines for the M1s. Based on the 635 CSi coupe engine, they added a twenty-four-valve head used in IMSA racing in the United States. In the M1, the 3.5-liter six developed 227 horsepower at 6,500 rpm and 243 lb-ft of torque at 5,000 rpm.

Lamborghini was contacted to design and develop the M1 chassis and then build the cars. The Italian automaker fell on difficult times, however, and the project had to be pulled from it. As a result, the frames were made by an Italian firm and the bodywork added by Italdesign, with final assembly done in Germany by Bauer.

By twenty-first-century standards, the M1 interior doesn't look particularly aggressive. Compared to the often overly firm seats in some supercars, the M1's are downright comfy, and they get the job done in hard cornering. The sound of the classic BMW six behind the cockpit at speed is music to the ears of an enthusiast.

Giampaolo Dallara, who had done the Miura and Countach. Dallara still builds IRL race cars today.

Lamborghini was also meant to build the cars, but here came the first hiccup. It may be owned and supported by Volkswagen via Audi these days, but in the 1970s and 1980s, Lamborghini was like a ship in heavy seas, sliding deep into a financial trough before being saved, only to slip away again. It was during one of those deep slips that BMW had to pull the M1 project from the Italian automaker. A firm called Marchesi welded the frames together, while Trasformazione Italiana Resina made the fiberglass bodies. Italdesign fitted the bodies to the frames. Each car was then shipped to the well-known German coach-building company, Bauer, for final assembly and the addition of the BMW Motorsport engine and ZF gearbox.

It was a promising package. Motorspot tuned the 3.5-liter incline-six to 277 horsepower at 6,500 rpm and 243 lb-ft of torque at 5,000 rpm.

Starting with the engine from the 635 CSi coupe, they added twenty-four-valve twin-cam heads designed for use in the 635 coupes that raced in the US International Motor Sports Association (IMSA) series in 1974, along with Kugelfischer-Bosch fuel injection and electronic ignition with no distributor, something common today but new then.

Set longitudinally behind the cockpit and bolted to the ZF five-speed transmission, the BMW six motivated the M1 to 60 miles per hour in, by BMW's timing, 5.3 seconds, on its way to a top speed of 162 miles per hour . . . not impressive today, perhaps, but enough to turn heads in 1978; the car was launched at the Paris Auto Show in October of that year.

The rest of the chassis was de rigueur for the time: upper and lower A-arm suspensions (the rears with radius rods), nonassisted rack-and-pinion steering, vented disc brakes, and alloy wheels, here fitted with the famous Pirelli P7 tires.

Enter hiccup number two. With all this fussing around to get the car built, its original purpose was slipping away as the homologation rules were revised. The great plan to take on Porsche with an 850-horsepower turbocharged Group 5 M1 fell apart. There were M1s built for Le Mans—the most famous being painted as part of a BMW art project by Andy Warhol—but they were never a threat to Porsche.

To garner some notoriety with the M1s, BMW created the Procar series, pitting Grand Prix drivers against each other in a 470-horsepower Group 4 M1s as curtain raisers to many Grand Prix in 1979 and 1980. This was the European version of International Race of Champions (IROC) and just as competitive.

The most prominent of the M1's racing honors came in the United States, where David Coward and Kenper Miller won the 1981 IMSA GTO Championship. By that time BMW was off on another racing tack, building Formula 1 turbo engines for Brabham.

In the end, BMW built around four hundred M1 street machines and an additional forty-five to fifty as race cars. Today, a nice one brings $200,000. What few M1s were originally imported for sale in the States were then certified by independent firms to meet safety and emissions rules, bumping the price back then to around $115,000.

BMW has M1s in its collection in Munich and brought one for our day in the southern German countryside. It's still a great-looking automobile—arguably Giugiaro's best wedge sports machine. We're now used to seeing engines hidden under styled covers and panels, so it's nice to lift the M1's large rear panel, look in, and see a naked engine. Mind you, it's a stretch, well forward in the bay, but there is BMW's beloved straight-six, injectors under a small air filter on the left, headers snaking down and forward to the right. How beautifully uncomplicated.

The M1 interior looks so period, with a squared-off instrument binnacle containing little gauges. The steering wheel feels slim and looks to modern airbag eyes to be a bit lethal, but it is the perfect shape for the racer's hands at a nine-and-three grip.

While the Recaro seats don't have the aggressive visual appeal of those in today's performance cars, they are arguably better. For some reason or other, modern interior designers seem to believe a sports car seat needs to be very firm and high-sided to hold you in. By contrast, the M1's is surprisingly soft, like sitting on a comfy cushion. While that may not be the best on those occasions when you're cornering hard, it certainly makes this one of the more comfortable-riding high-performance cars of its day.

Shifting the M1 is a vintage matter, with a little hitch in the neutral slot as you change up or down, combined with double clutching to ease the process. That was expected then, as automakers were less than a decade into building production midengine cars in any numbers. Thus, getting that shift pattern duplicated way back there behind the engine involved a lot of levers and linkages, including bushings or whatever to keep them from being noisy.

But rhythm is the key, and once you get into the rhythm of shifting, the M1 is quite easy.

When BMW could no longer get enough speed from its 3.5-liter CSi coupes to beat its German rival Porsche on the racetrack, the Munich firm decided to create the midengine M1 to uphold the company's honor in competition. M1s were raced by Grand Prix drivers in the Procar series, and one won the US IMSA GTO championship in 1981. Easily the most memorable M1 race car, however, was this Le Mans entry. Its fame came because it had been painted by Andy Warhol as part of a BMW art project.

Famed Italian chassis engineer Giampaolo Dallara was responsible for the underpinnings of the M1. The BMW M1's philosophical predecessor was designer Paul Bracq's 1972 Turbo concept car, done to commemorate the Munich Olympics. Creation of the final production M1's design was given to Italian Giorgetto Giugiaro, though hints of Bracq's Turbo shape remain in the BMW.

Another odd thing: back then, German automobiles insisted on putting first gear in their five-speed transmissions in the lower left slot, down below reverse. This was a racing tradition because first was a launch gear and it was more logical to put the four working gears in the H pattern.

Nowadays, most cars are developed so all the driver inputs are well balanced—steering, brake effort, throttle. The M1's non-assisted steering, by contrast, is a bit heavy at low speed but lightens nicely as speed rises and is about right and nicely direct at speed. Brake and clutch efforts seem quite logical for a twenty-five-year-old high-performance car.

Too bad timing and the times seemed to trip up the M1 project. It had the right styling pedigree, one of the best chassis designers in the business, and the beloved BMW straight-six. Even after two and a half decades, it stands out in the exotic car crowd like an honored, dignified middle-aged statesman, still quite athletic . . . and with an excellent tailor.

A supercar among BMW's many super cars.

The Group B Connection: 1983–1991

Two great Jaguars: the D-Type (in the back) and the XJ220.
Designer Keith Helfet gets credit for the supercar's sleek exterior.
His goal was to create a shape that would contain and cool a
500-horsepower engine, be aerodynamically stable at 220 miles
per hour, and look like a Jaguar.

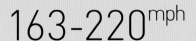

163-220^{mph}

While the midengine 288 GTO has its own fan club, the original
front-engine GTO is a legend. The 250 GTO was based on the Testarossa
sports racers that had won so many victories for Ferrari, a tradition the 250
GTO continued by winning the FIA's Manufacturer's GT Championship in 1962,
1963, and 1964.

Ferrari 288 GTO 1983

189^{mph}

Got 10 million bucks? That's the least you'd need to buy an original 1962 Ferrari 250 GTO. Probably more, but probably enough to buy all the original Porsche GTOs still worth owning.

Or for about $400,000, you could have a 288 GTO, a limited-edition Ferrari that will outperform the other two. One that turbos to 60 in 4.9 seconds, to 125 in 15.2, and doesn't want to stop until you hit 189 miles per hour.

First, a bit of history. Based on the Testa Rossa sports racing cars that almost dominated international racing in the late 1950s, the 250 GTO (for *Gran Turismo Omologato*) won the Federation Internationale de l'Automobile (FIA) Manufacturer's GT Championship in 1962, 1963, and 1964, plus the hearts of so many of us. With arguably the most beautiful front-engine GT shape ever created, strong and reliable V-12 power, and a predictable, unbreakable chassis, GTOs rose to the top of the Ferrari list—hence their eight-figure price tags.

All Pontiac did was steal the name. And then Ferrari reused it.

As with Porsche's 959 and Jaguar's XJ220, the Ferrari 288 GTO was aimed directly at the FIA's Group B race program. Like the Jaguar, the new GTO became a historical postscript to that short-lived racing class, but it is still highly regarded among Ferrari owners.

Ferrari said it would build the GTO in 1983. Unlike Jaguar, which created entirely new automobiles for Group B, Ferrari based its machine on an established production car, the 308. By doing this, the automaker was starting with a known commodity, and there could be marketing spillover to support sales of the normal production car.

Admittedly, there wasn't much of the 308 left when the 288 GTO was finished. The standard semimonocoque structure had been replaced by a lighter steel tube frame and separate body panels. While it had the same upper and lower A-arm suspension as the 308, the GTO's arms were of tubular steel and the shocks and springs firmer for the added engine power.

Pininfarina, which designed the 308, did the reworking for the GTO, buffing up the original design with muscleman fender flares, adding serious spoilers front and rear, and plugging in four huge driving lights. A holdover from the original 250 GTO was a set of slanted air vents, which had been behind the front wheels on the 250 GTO but vented air from the rear brakes in the new car.

Even the bodywork material was new and was kept lighter for the race/ road car. Using steel only for the doors, Ferrari molded fiberglass for the major body panels (as it had for the very first small run of 308s). Kevlar was used for the hood, combined with carbon fiber for the roof and with Nomex and aluminum honeycomb for the firewall.

In perhaps the most fundamental change from the production car, the engine was turned from its lateral side-to-side placement in the 308 to a longitudinal fore-aft layout for the 288 GTO. In swinging the drivetrain around, Ferrari had to extend the 308 wheelbase by 4.3 inches, making it 96.5. This rearrangement meant it could use conventional race car engine/transmission layout, which would be crucial for quick gear ratio changes. Also, one of the turbos didn't have to be fitted in between the engine and firewall.

Ah yes, turbos. Because to get the sort of power one needed for an exotic car in the early 1980s, adding turbos and intercoolers was a logical solution. In the GTO's case, it was a pair of IHI-made turbos, one per side and kept small to lower inertia so they could spin up easily and minimize lag. Using turbos meant Ferrari had to lower the V-8's displacement by a "turbo factor" of 1.4—the FIA's mathematically calculated advantage of the little exhaust-driven superchargers; the result had to be less than 4,000cc.

While the GTO's twin-cam, thirty-two-valve V-8 was derived from that in the 308, it understandably had major internal changes to allow for the turbocharging and increased horsepower. The compression rate was lowered to 7.6:1 and the boost set at 14.5 psi, and it got new Weber-Marelli electronics for fuel injection and ignition.

Originally, Ferrari intended to race the 288 GTO in the FIA's Group B class against cars such as Porsche's 959 and Jaguar's XJ220. The class was short-lived, and this modern GTO was never raced in anger.

Ferrari lowered the displacement of the 308's V-8 to 2,855cc so it could then turbocharge the engine. With a compression ratio of 7.6:1 and running 14.5 psi of boost, the 288 GTO engine was rated at 400 horsepower at 7,000 rpm and 366 lb-ft of torque at 3,800 rpm.

To get the 288 GTO ready for racing, Pininfarina squared its shoulders with wide fender flares and added spoilers under the nose and atop the tail. Four driving lights were plugged into the grille.

While Ferrari's 308 has a semimonocoque structure, the 288 GTO uses a steel-tube frame with separate body panels. The doors have steel skins, with the major body panels in fiberglass, while some components are in Kevlar, carbon fiber, and other lightweight materials.

The seat style, the stitching, the steering wheel, and the gated shifter are all pure mid-1980s Ferrari style.

On the dyno, the results were impressive: 400 horsepower at 7,000 rpm and 366 lb-ft of torque at 3,800. With the help of a five-speed gearbox and a curb weight of just 2,555 pounds, *Road & Track* managed to get a 288 GTO to 60 miles per hour in 4.9 seconds. Terminal velocity topped out at 189 miles per hour.

Sadly, while Ferrari had no trouble selling the 287 Ferrari 288 GTOs it made, the cars were never raced in anger. Rule changes and corporate priorities kept the GTOs out of the race garage.

And then one day in 1987, we were working at Ferrari's Fiorano test track on a story for *Road & Track* when a rather amazing-looking 288 showed up in the pits. Was it a competition car? Sure looked the part, but Ferrari told me its correct name was the GTO Evoluzione.

Hmm, we'd never heard of it. Ferrari went on to explain that it was one of three such cars made, and it used turbo V-8s that were a development of the GTO. Ferrari had developed two engines, both with boost pressure set at 24.7 psi. The first, called the 114 CR, pumped out 530 horsepower while the second, the F114 CK, produced 650 horsepower.

Hmm again. They told us the more powerful GTO/E got to 60 miles per hour in about 3.5 seconds with a terminal speed of 240 miles per hour.

There were other changes too, such as a tube-frame chassis that was reinforced with carbon fiber—the latter material used for a body that created a machine 40 percent lighter but three times stiffer than a GTO.

A test driver did any number of laps with the GTO/E, its right seat filled with instrumentation, hanging the back end out for me. Wow. And then the car was gone.

What I didn't realize until later was that I was looking at one of the prototypes for the soon-to-be-introduced F40. Using the GTO/E as a yardstick against Porsche's 959, Ferrari was developing the first of its limited-run exotic cars.

By the early 1990s, when the 512M version of the Testarossa was produced, the big midengine machine had been developed into a strong handler that could generate mid-0.9g handling around a skid pad. Here, World Driving Champion Phil Hill plays with a 512TR on a damp racetrack.

Ferrari Testarossa
1984

185^{mph}

"At 175 miles per hour or so—this feels like cruising speed—the car is stable, the handling feels taut and secure. Pushing up over 180 miles per hour, you begin to detect just a hint of lightness in the front end and there is a dramatic awareness that you are going incredibly fast. With sufficient room, the powerplant eventually nudges the redline in fifth gear and you flash past the [racetrack's] lighted sign that shows your speed: 185 miles per hour. Back off a little, the speed drops into the 170s, and the sure-footedness returns as though you were cruising along the Interstate." —*Road & Track*

Been there, witnessed that from the passenger's seat.

The preceding is a description of what it was like to peg the tach—world champion Phil Hill driving—in a Ferrari Testarossa. This was more than thirty years ago at a *Road & Track* World's Fastest Cars event, back before exotic cars got the wind tunnel time and

aerodynamic attention they receive today, so very high speeds could get your attention.

Despite that speed and the fact that it's difficult to be lukewarm about any Ferrari, that word—lukewarm—is how many of us remember the Testarossa, or TR.

Mind you, Ferrari sold a fair number of TRs, and they were solid, reliable automobiles, but they weren't exactly enchanting. Sorry to burst any bubbles.

They also weren't that quick in their initial iteration, *Road & Track* needing 6.2 seconds to get a US-spec first-generation TR to 60 miles per hour. (European TRs were almost a second quicker, back when emissions controls often caused a dramatic difference in Euro versus US horsepower. This disparity disappeared in the early 1990s as government rules between the two continents became more similar and techniques, particularly electronics, improved.)

71

While the width of the Testarossa might be a problem when driving, it brings the benefit of a broad interior. Once over the wide doorsill, you can settle into very comfortable, leather-upholstered seats.

Testarossa means "red head" in Italian, a reference to the color of the engine cylinder heads of this midengine Ferrari exotic. The name was inherited from the highly successful Testa Rossa sports racing cars of the late 1950s and early 1960s.

Like the Boxer, the Testarossa is propelled by a flat-twelve engine. In its original form, it displaced 4.9 liters and generated 380 horsepower and 354 lb-ft of torque. With the 512M version in 1992, horsepower increased to 421.

There is no denying the Testarossa is fun to drive. Power from the flat-twelve behind your back is eager to punch you down the straight and to the next corner, and you can guide a Testarossa with surprising ease, given its potential . . . and its size.

If you have ever driven on the roads in the hills behind Ferrari's factory in Maranello, Italy, you would understand why the automaker's cars handle so well. They have nicely paved two-laners that twist over the hilltops, down into valleys, and through towns like Serramazzoni. It isn't unusual to see a Ferrari test car or camouflaged prototype in the hills.

Ferrari Testarossa 1984
73

On these roads, you would find the TR drives "narrower" than it is; it is easier to maneuver than expected, given its 77.6-inch width (0.6 inch wider than a Range Rover.) But you are always aware of its big backside, something the limited visibility to the rear reminds you of every time you check the mirrors.

Many bicycle riders cycle on those winding roads behind Maranello and, with the TR, you are aware of the potential for sweeping one into the weeds.

And then there is the car's exterior styling.

Ferrari's first midengine car, the Berlinetta Boxer, is a better automobile than the Lamborghini Countach and, some would argue, just as attractive in its own not-in-your-face way. The styling of the Countach, however, so captured everyone's attention that it seemed the Boxer's successor would have to make a strong statement.

It did. When Ferrari distributed the first photos of the Pininfarina-designed Testarossa, we were all amazed. It was a big-hipped shape, the front smooth with pop-up headlamps, but what is that along the side? Slicing into the rear duct were five . . . what are they? Some critics named them strakes, while others called them a stack of little fins. Was it a cheese grater? In any case, it got our attention, as did five similar lines across the back of the car.

Technically, the TR has a lot going for it. Out back is the flat-twelve, forty-eight-valve, twin-cam boxer engine displacing 4.9 liters and producing 380 horsepower at 6,300 rpm and 354 lb-ft of torque at 4,500 rpm—plus red cylinder heads, as Testarossa means "red head." Behind it is a five-speed manual gearbox that is shifted up front through a traditional, handsome shifter gate.

Classic chassis specs: upper and lower A-arm coil-spring and tube-shock suspensions, rack-and-pinion steering, and vented disc brakes (12.2 inches) all around. The Goodyear tires are on alloy wheels with the famous five-spoke Daytona pattern.

While you must be aware of the TR's width when driving, its benefit is a generous cockpit, at least in width, as the design's low height makes it a bit tight for those over 6 feet tall. Once you have climbed in over the wide sill, the interior's leather ambiance and scent are (at least when new) enticing, the gauges classic.

Exotic cars had lost much of their fussiness when the first TR was introduced in 1984. Clutches were no longer too heavy, and the threat of stalling the car on launch (how embarrassing!) was minimal. Steering effort had leveled out, and the air conditioning worked on cool days.

By 1992, the Testarossa was a bit long in the tooth and upgraded to the 512M. Displacement didn't change, but horsepower certainly did, up to 421 at 6,750 rpm. Torque was up only 6 lb-ft, but the curve was broader, adding to driving flexibility. The brakes were better, the tires bigger, and the car would circle a skid pad in the mid-0.9 g area. It got off the line smartly at just 4.7 seconds to 60 miles per hour with a top speed of 192.

Pininfarina designs tend to be well known because they establish a long-term trend rather than follow a fad, but the TR seemed to do the latter. You won't find the car on many (if any) "Best Ferrari" lists. When its successor, the 550 Maranello, turned out to have a front-engine, rear-drive layout, no one was disappointed.

Ferrari champion Phil Hill wrote of the 512TR: "At 175 miles per hour or so—this feels like cruising speed—the car is stable, the handling feels taut and secure. Pushing up over 180 miles per hour, you begin to detect just a hint of lightness in the front end, and there is a dramatic awareness that you are going incredibly quickly."

If there is one overriding feeling when driving a Testarossa, it is that you are guiding a very wide car. This is, of course, a result of putting a large, flat-twelve powerplant amidships and building a supercar around that layout.

At the debut of the F40, Enzo Ferrari explained the car's name: "It is forty years since the first Ferrari left the factory. On March 12, 1947, the 125 S was presented to the public for the first time. On May 25 of the same year, Franco Cortese won the Rome Grand Prix." As this photo shows, there are several ducts on the F40 to add cool air or relieve hot air from the supercar, which is a problem in high-horsepower supercars that need to be as adept at speed on a track as at a crawl on the street.

Ferrari F40
1987

201^{mph}

July 12, 1987, Maranello, Italy

When Enzo Ferrari arrived, the place went nuts. Journalists—normally cool, reserved, standoffish—stood to applaud and cheer the Pope of Maranello. Photographers rushed toward the table behind which Ferrari was to sit, and the repeated flashes from their strobes gave the darkened hall a strange disco feeling.

"Please, no flash photos," the public-address system asked. Now, there were more than ever.

They had to be kidding. This was Enzo Ferrari. Predictable pandemonium.

I was sitting well up in back in Modena's small Centro Civica auditorium in Ferrari's museum, watching all this unfold like a strange comic opera. It was grand entertainment. You have to understand that we attend numerous press conferences each year, and in most cases, those putting on the event are just happy to have journalists attend. This, on the other hand, was a command performance.

The invitation was almost informal: "If you happen to be in Modena this particular day, you might be interested in seeing this new car." We journalists pitch some engraved press invitations in the trash, but this quiet invite from Ferrari was enough to cause us to drop everything to be in Modena on the appointed day. We knew it had something to do with the rumored new model dubbed the "Le Mans," successor to the 288 GTO, but beyond that, it was a mystery.

The bar of Modena's Fini Hotel the night before the press conference had sounded like a session of the United Nations, with many languages heard all around. No one was about to miss this evening.

Ferrari added to the mystery the next day when we entered the hall to find a lovely draped shape sitting under spotlights. Starting time was promised at 11 a.m., but then rumors were that Mr. Ferrari would be ten minutes late. Despite the heat inside the building, no one minded. Finally, a side door opened, adding a corner of bright light to the dimly lit hall. Mr. Ferrari, now eighty-nine years old, was helped to the table. That's when the place went nuts.

The Commendatore began:

> On July 6 last year, I asked my research department to look into the feasibility of building an exceptionally powerful sports car incorporating the very latest developments in engine and assembly technology. Only six days later, the directors gave the project their blessing, and now, barely a year later, the finished car stands before you.
>
> It is forty years since the first Ferrari left the factory. On March 12, 1947, the 125 S was presented to the public for the first time. On May 25 of the same year, Franco Cortese won the Rome Grand Prix, driving a Ferrari. Now, forty years later, the Ferrari F40 demonstrates that Ferrari is still a byword for technological excellence and exceptional performance.

Everything went nuts again as the red drape was pulled from the F40, Ferrari's response to Porsche's 959.

There followed speeches by Giovanni Razelli, president of Ferrari; Leonardo Fioravanti, who led research and design for Pininfarina; and the man who had a great deal to do with the engineering development of not only the F40 but also the GTO and Evoluzione, Nicola Materazzi.

Razelli acknowledged that the F40 was a direct product of racing, using what Ferrari could learn on the track to test ideas for production cars.

Enzo Ferrari was the real star the day the F40 was presented to the public. The famous man is off in the shadows to the right, a photographer taking aim at him. It was a hot, sweaty day in the presentation auditorium, but no one cared, because they had a chance to see Ferrari . . . and the F40.

Ferrari forsook aluminum for the F40's body, going instead with carbon fiber and Kevlar. That big front deck weighs a mere 39.6 pounds, while the huge rear engine cover tips the scale at just 48.5 pounds. Each door is only 3.3 pounds, complete with hardware.

Ferrari F40 1987
79

Fioravanti explained the design development of the F40. Naturally, Pininfarina's famous wind tunnel was used to develop the shape of the F40, which has a coefficient of drag of around 0.34, combined with low lift figures front and rear.

But forget all the facts and figures, because first of all the car is damn exciting. In describing the F40, Fioravanti, who is a fan of vintage automobiles, hearkened back to Ferraris of the 1960s. Consider the shapes of the F40 and 959, and then remember the form of the two companies' race cars of 1960. Porsches featured round, organic shapes, while Ferraris had that specific Italian look, sleek and racy. The descriptions still apply.

In front, the F40 has one major opening, which doubles as an intake for both the oil cooler and the optional air conditioning. Outboard of this are brake-cooling holes. The NACA ducts on the hood provide the breeze to cool the interior, while the pair on each flank of the car cools the brakes and engine compartment. Behind each flared front

Ferrari was developing the F40 near the end of the era when turbochargers were often used on supercars, a technology only Porsche has continued to use. Ferrari tacked two turbos on the F40's 2.9-liter V-8 with air-to-air intercoolers and got the engine to whoosh 478 horsepower at 7,000 rpm and 425 lb-ft of torque at 4,000 rpm.

Ferrari's contemporary competition was Porsche's 959, and the two had similar (high) performance. The 2,420-pound Ferrari clocked a 0–60 time of 3.8 seconds, topping out just over 201 miles per hour, about what the super Porsche could achieve. Where the Ferrari appeals very much to the heart, the Porsche goes for the brain.

fender is a duct to release hot air. That big rear wing is there to add downforce, while the back window is specifically shaped to direct flow onto the wing.

When doing the interior, function was the point of Pininfarina's design. There was a choice of three types of seats for the new owner, who had to visit the factory or his dealer to be fitted for one. The plastic composite seats weigh about 3 pounds each. They're finished in red and feature a full set of competition belts.

Although air conditioning was an option, power windows were not. In fact, there weren't even wind-up windows. Ferrari chose sliding windows to fit the car's functional and traditional themes, though roll-up windows were added later. Similarly Spartan, the pedals are bare metal and drilled. Set in the dashboard is a classic-looking Ferrari instrument panel fitted with a tachometer, speedometer, fuel gauge, and a dial for oil pressure. In contrast to the red seats and door panels, the rest of the interior is finished in flat black.

With previous Ferraris, you would have expected to hear about panel beaters pounding out aluminum F40 bodies on forms. By 1987, Ferrari had learned to manufacture bodies of carbon fiber, Kevlar, and resin. The single-piece front deck weighs in at 39.6 pounds, with the rear one a mere 48.5 pounds, including the wing. The doors, complete with hardware, are 3.3 pounds.

Carbon fiber, Kevlar, and other advanced materials were also used in the F40's chassis. Things began conventionally enough with a steel-tube frame, but the frame was reinforced with panels molded in plastic and glued in place. A few sections, such as the front well for the spare tire, looked like plastic sculptings. Naturally, critical areas, such as the engine and suspension mounting points, were metal. Fitted inside the body just ahead of each rear wheelwell was one 15.8-gallon fuel cell per side. Competition filler pipe and cap were available on request.

Ferrari claimed the "plastic car" was several hundred pounds lighter but four or five times stiffer than a traditional metal/frame body design.

Both the front and rear suspension of the F40 were laid out with unequal-length A-arms, coil springs, Koni shocks, and anti-roll bars. The change for Ferrari was to add adjustable ride height, the F40 settling automatically at high speeds—down 0.8 inch above 74 miles per hour—and having the option of adding an extra inch of ground clearance to get over such obstacles as driveway ramps. F40 tires are Z-rated Michelins, measuring 245/45ZR-17 at the front and 335/35AR-17 in back.

Tucked inside all that rubber are 13-inch-diameter disc brakes. To keep weight down, both the four-piston calipers and discs are aluminum, the latter with cast-iron braking surfaces set in the discs. No power assist here, and no ABS, thank you; that just was not in the tradition of these cars.

Ferrari also explained that the brake system was up to racing standards, " . . . and this means a system which can be used for competition purposes if necessary." It isn't surprising, then, to discover the rack-and-pinion steering is also unassisted.

Ferrari developed the 288 GTO Evolucione's twin-turbo V-8 for the F40. Displacement was increased to 2,936cc, and bigger water-cooled IHI turbos were added, with the maximum pressure set at just over 15 psi blowing through a pair of air-to-air intercoolers. Weber-Marelli designed sequential, phased electronic fuel injection for the F40, which would prove to be a particular benefit to US Ferrari owners several years later.

This mechanical stuff was all very interesting, but most exciting is to see the V-8 set fore-aft under the louvered rear window. One look in the tail section and there is no doubt this car is a race car waiting to get down to business. And it has the power, with 478 horsepower at 7,000 rpm and 425 lb-ft of torque at 4,000.

F40 owners had the choice of two different gearboxes, but it wasn't the usual question of manual or automatic. In the Ferrari, you could specify either a five-speed with synchromesh or, for those who wanted to feel they were in a real race car, the same gearbox without synchros.

Three types of seats were offered in the F40, and the driver had to be fitted for the proper seat at the factory or by a dealer. Though the interior can be somewhat noisy, there are amenities, such as optional air conditioning and roll-up windows.

This is a US version of the F40. Ferrari did an excellent job of reworking the Pininfarina body to accommodate the "safety" bumpers required by the US government. Engineers had to deactivate the adjustable suspension on the US F40s to meet the bumper-height rules.

Ferrari's F40 production line was a slow-paced affair . . . and a long-running one. Owners criticized Ferrari for building too many F40s (1,315), diluting the pride (and profit) of ownership.

One thing the F40 designed for was four-wheel drive à la Porsche 959. Ferrari built a prototype with drive to all wheels but said it preferred to keep the F40 in the more traditional style.

Not that owners of the Ferrari had to apologize to those with an original 959. Weighing in at about 2,420 pounds, the F40 got to 60 miles per hour in 3.8 seconds with a top speed of 201 miles per hour, the times all within a tick or two of the 959.

That should be enough for most of us . . .

Where Porsche's 959 was a technological showcase, the F40 was more a display of Italian brio, speed, and beauty. The brain opted for the 959, while the heart went for the F40.

At the time the F40 was literally unveiled on that hot day in July 1987, it was promised there would be a US version, though the process took three years. Building F40s at the rate of one to two per day, Ferrari was still at it three years later and, in that time, a great deal had been certifying supercars for the United States and happily applied to the F40.

Externally, there was little difference for the US edition. Small black strips above and below the grille act as impact surfaces for the 2.5-mile-per-hour bumper crash regulations. Ferrari deactivated the adjustable suspension on cars for the States rather than design a bumper tall enough to be effective through its upper and lower heights. There was also a rear safety bumper, and like the front, it had added internal structure to meet the rules. Also, outside were numerous small lighting changes, including a high-mounted rear brake light.

Inside, the only major change was a motorized "mouse" seat/shoulder belt to meet passive restraint regulations.

Thanks to that Weber-Marelli electronic fuel-injection system and the latest catalytic converters, Ferrari engineers were able to Americanize the twin-turbo V-8 without a major power loss, still claiming 478 horsepower and 425 lb-ft of torque.

In July 1987, Ferrari hinted it would cap F40 production at four hundred cars, but the model proved so popular (and profitable) that it just kept building them, finally completing 1,315 cars . . . and angering many of the early owners who were looking for rarity in a supercar. Some had paid up to $1.5 million for the F40.

Ferrari also eventually got around to having a race version of the F40 built. It had Michelotto in Padova do the conversion, which was called the F40 LM. The interior was stripped still more and a digital dash installed, while the exterior was tuned to competition, replacing the fixed rear wing with one that could be adjusted for downforce.

Power was boosted to 720 horsepower with the potential for 900 in qualifying sessions, a trick that was common in the turbo days. Formula 1 driver Jean Alesi drove an F40 LM in its racing debut at the Laguna Seca's IMSA race in October 1989, and he led the race until his tires let him down and he finished third. A pair of F40 LMs was campaigned in IMSA in 1990 and had four podium finishes, though never a win.

Racing F40s were successful in the mid-1990s in the European BPR Championship and in a Japanese series, with some victories against newer, more completely developed cars. Minimal development also hurt the F40 LMs' chances in endurance races, such as the 24 Hours of Le Mans.

This lack of racing success shouldn't take away from the aura of the F40. At the time, it and the Porsche 959 were the supercars. Drive an F40 with its barely tamed race car demeanor—with mechanical sounds reverberating through every piece as you rap up through the gears—and you discover a delightful hard edge that is lost, for better or worse, in the exotic cars of the twenty-first century.

At the famed Nardo high-speed track in southern Italy, this Jaguar supercar managed 219 miles per hour, close enough to earn the car its XJ220 moniker. Like Porsche's 959 and Ferrari's 288 GTO, the XJ220 was originally meant to be raced under the FIA's Group B rules, which were later rescinded. Jaguar hand-built the XJ220s in a converted mill in Bloxham, not far from Banbury, England. After the XJ220 production run was finished, the factory was turned over to Aston Martin DB-7 production.

Jaguar XJ220
1988

220^{mph}

"That was one of the scariest things I've ever done in a car."

"Really? Why?"

Davy Jones, ace Jaguar sports car driver (and future Le Mans winner) explained the dangers of holding the Jaguar XJ220 on the 8-mile, high-speed banking near Nardo, Italy, as we went for top speed. He mentioned the single-rail steel barrier and how, at high speed, we'd be long gone before anyone even knew we'd crashed.

Ah, the bliss of ignorance on my part. I had thoroughly enjoyed the ride, fitted with fire suit and helmet and strapped into the long Jaguar. We'd seen 216–217 on the speedometer, subtracted for speedo error, added a mile-per-hour for tire scrub, and been able to claim a top speed just one tick shy of the big cat's namesake number: 220 miles per hour. Close enough.

Appropriate, too, for while Jaguar's famous sports car, the XK120, was meant to go 120 miles per hour, its newest was aimed at 220.

What made the speed so deceptive was the nature of the XJ220. Where other supercars had a slightly hard edge or a sense of a muscle-taut sprinter just out of the starting blocks, the Jaguar did not.

That image began with the exterior shape, which was sleek and flowing compared to other exotics, an obvious 1990s extension of the 1950s Jaguar D-Type and 1960s XJ13 race cars. A member of the family.

Inside many high-speed exotic cars, there's a feeling of a tuned-down race car with hard seats and minimal luxury. Not in the Jag, its dashboard broad and like a production car, leather everywhere, and seats looking downright overstuffed compared to those in a Ferrari or Lamborghini. Knowing the low-roof design created a dark, bunker-like effect inside, the car had added a glass roof so the interior was light and welcoming, nowhere near as threatening as some supercars.

Quite simply, Jaguar's XJ220 was the limo of the supercar set.

Credit for the idea of the XJ220 goes to Jaguar's then-director of product engineering, Jim Randall, who wanted to build a Group B machine—à la Porsche's 959—based on the 1984 Jaguar race car. Specifications included V-12 power and four-wheel drive. There were no production intentions, and for three years, a group of Jaguar employees, who called themselves "The Saturday Club," worked on the 220 during off-hours. Outside suppliers donated many of the materials needed.

Keith Helfet of Jaguar Design did the exterior design of the 220, which had to be street legal and cool a 500-plus-horsepower engine at a sustained 200 miles per hour. Manufacturing was kept in mind, just in case . . .

Nick Hull created the interior, trying to capture the wraparound feeling of the D-Type car, but with air conditioning, leather upholstery, power windows, and other driver amenities. What the XJ220 brief did not include were such driving niceties as power steering, cruise control, and automatic transmission. Back then, they didn't seem right in an exotic car.

Shown to the public at the 1988 Birmingham show, the 220 was so well received that production was approved. Jag had neither the time nor the facilities to build the limited-production supercar, so the job went to JaguarSport—the joint project of Jaguar and Tom Walkinshaw Racing (TWR) meant to take Jag racing and do just this sort of low-volume, high-performance project.

Getting the XJ200 to production necessitated changes. Most significant was submitting the 3.5-liter turbo V-6 from the XJR-10 and -11 race cars for the V-12 in the show car and using rear- instead of all-wheel drive.

One of the first aims was to trim weight from the 3,700-pound show car XJ220. TWR got the heft down to just over 3,000 pounds, a hunk gone when it shortened the car's overall length by 10 inches, 8 inches of it gone from the wheelbase—a move made possible when the V-12 went, in favor of the V-6.

The XJ220's structure was based on panels of aluminum honeycomb sandwiched between aluminum sheets. Exterior bodywork was done

If there was a technical downside to the XJ220, it was its engine. Originally the engine was meant to have V-12 power, but it ended up with a 3.5-liter twin-turbo V-6. It was powerful enough, with 542 horsepower at 7,000 rpm and 475 lb-ft of torque at 4,500 rpm, but it sounded as though it belonged in a super tractor, not a supercar.

Jaguar's XJ220 started as an after-hours "what if?" project headed by Jim Randall, the highly regarded then-director of product engineering for the automaker. A group of Jaguar employees called "the Saturday club" donated their time for the car's early development. And though the XJ220 had plenty of speed of style, it fell victim to the world economy. Debuted in 1988 and approved for production the next year, the car was priced at $660,000 and required deposits of almost $100,000. Between order time and delivery two years later, the exotic car market fizzled. Jaguar ended up in court with supercar buyers.

Davy Jones slides the big Jag through a handling course. The XJ220's handling was developed at England's Donington circuit and Germany's Nürburgring, where it held the production car lap record. High-speed testing was conducted at Nardo in Italy and MIRA and Millbrook in the UK.

There is none of the hard-edge, almost-race-car feeling inside the XJ220 that you get in some supercars. The interior is pure Jaguar, from the comfy leather-covered seats to the instrumentation to the dashboard. To keep the interior from feeling low and dark, engineers sensibly gave the XJ220 a glass roof.

In the mid-1960s, Jaguar came close to reentering international competition with the mid-engine XJ13 race car. Powered by a 500-plus-horsepower V-12, the XJ13 was designed in 1964, but slow development put it behind the times, and Jag killed the project. The XJ13 was kept a secret until 1973. The car is now in Jaguar's museum in England.

in aluminum by the famous firm of Abbey Panels in Coventry, England, while hidden inside was an integral steel roll cage.

The 220 body wasn't just pretty and safe; it was also aerodynamically efficient. The coefficient of drag was 0.36, and it had 600 pounds of downforce. At the front was a fixed lower splitter to get the air under the car, where it flowed through channels to the rear venturi outlets, aiding downforce. Air through the nose grille cooled the engine and air conditioning condenser, then vented out of the top of the nose. At the back was a full-width spoiler.

The front and rear suspension designs were unequal-length upper and lower A-arms. Coil springs and Bilstein shocks were mounted inboard and operated through rocker arms. Bridgestone supplied the tires, Speedline the single-piece alloy wheels, and AP the non-ABS vented disc brakes.

Fuel was held in a racing fuel cell between the cockpit and engine compartments.

Looking almost diminutive in the back of the XJ220 was the 3.5-liter V-6 twin-turbo with dual overhead cams, four valves, and some 25 psi of maximum boost. The dry sump V-6 looked like a race engine in street dress with an air conditioning compressor and alternator. Companion to the V-6 was a five-speed manual transaxle, which had a viscous limited-slip differential.

Here's how the XJ220's horsepower and torque stacked up to the Italian competition of the day:

Jaguar XJ220	542 bhp @ 7,000 rpm	475 lb-ft @ 4,500 rpm
Lamborghini Diablo	492 bhp @ 7,000 rpm	429 lb-ft @ 5,200 rpm
Ferrari F40	478 bhp @ 7,000 rpm	427 lb-ft @ 4,000 rpm

And here are the results—using Jaguar's numbers—of the final performance tests against the F40 and Diablo:

	0–62 MILES PER HOUR	TOP SPEED
XJ220	3.85	212.3 mph (208 average)
Diablo	4.29	202.1 mph
F40	4.10	201.3 mph

High-speed XJ220 testing was done in Nardo, Italy, as well as at the MIRA and Millibrook testing facilities in England. Ride and handling work was completed at Donington Park Grand Prix Circuit in England at Nürburgring in Germany, where driver John Nielson set a new production-car lap record with the Jaguar.

With the XJ220's pedigree, good looks, and performance, it would be nice to say it led a loved, charmed life. That never happened, because the demise of the very thing that made it seem logical to build the car turned the XJ220 into a supercar orphan. The surging world economy (particularly in Japan) had fueled a dramatic rise in the demand and price of supercars in the late 1980s. When that world economy cooled dramatically in 1990, down went supercar values.

Jaguar approved the XJ220 in 1989 at a time when exotic car sales were storming along and priced it at $660,000 with a deposit of $94,000 on a signed contract. The company promised no more than 350 of the supercars would be built and buyers happily stepped up . . . including speculators planning on making a few bucks. This speculative disease, incidentally, affected the price of other exotic cars and valuable vintage cars around the world.

Bad timing on their part, as it turned out, because the exotic car market imploded in the two years between deposit time and when XJ220 production began in a converted old mill in Bloxham, not far from Banbury, England. Would-be buyers wanted out of their contracts on cars already worth less than the original agreed-upon value. Lawsuits ensued and dragged on until the spring of 1994, when the speculators were offered the chance to walk away from their contracts for one payment of $150,000.

There were also, of course, owners perfectly happy with XJ220, but by the time the dust cleared and production ended after 280–290 cars, Jaguar had a small warehouse of unsold XJ220s.

What to do?

Eventually, a batch was sold to a dealer as the possibility of certifying them for the United States improved. You can now find them in the States, certified and ready for the road.

But that didn't help Jaguar, and in summer 1993, the company thought it had a solution to some of the XJ220 problems. Jaguar had built a racing version called the XJ220C, which won its maiden race, a BRDC

(British Racing Drivers' Club) GT Challenge event. It was first thought an XJ220C had taken the Grand Touring Class in the 1993 24 Hours of Le Mans . . . until it was disqualified two weeks later.

Now the idea of racing returned, with a dozen XJ220s prepped by Walkinshaw for a US series called Past Masters. The idea was to match the XJ220s and former driving greats in a series of events at small US tracks, where the races could be televised easily.

The first race was at Indianapolis Raceway Park and it was embarrassing—like watching Carl Lewis in a gunnysack race or Jeff Gordon in a goofy bumper car. Suited to 200-plus miles per hour on Le Mans' Mulsanne Straight, the Jaguars looked like well-dressed ambitious fat men being forced around the tight, narrow road circuits and, often, into each other. Thankfully, the series was a failure.

It was also a sad, lasting remembrance of the XJ220 in many enthusiasts' minds.

Two great Jaguar shapes: D-Type and XJ220. While the D-Type has a long, honored history as a race car, the XJ220's competition aspirations as the XJ220C were limited, though it won its first race in the English BRDC GT championship.

Speedometer and tach needles are about pegged to the right as we rocket around the Nardo track just shy of 220 miles per hour. Driver Davy Jones—who later co-drove to victory in the 24 Hours of Le Mans—commented that the hot lap was scary. Luckily, the photographer was too busy (or too dim?) to realize this until Jones told him later.

The XJ220 deserves better. The cars have, arguably, the most elegant-yet-exciting styling of that era's supercars. And they drive the same way—nothing rough or rumbly, smooth controls, and a very nice fit and finish. It is a big car, and it takes a while to get used to moving that amount of real estate around, but it isn't intimidating.

Given the Jaguar's elegance, the feeling of 0–60 in 4.9 seconds seems a bit out of character, but not so much as the engine sound. For all the XJ220's beauty and grace, the V-6 sounds like a fight in a country bar. Ugly.

But only to the ears . . . to the eyes there is all the leather-wrapped luxury and elegance you expect from Jaguar . . . even at 219 miles per hour.

Porsche designed the 959 for Group B competition, which covered rally and road racing. Rules demanded an automaker build two hundred such machines, but put no limits on minimum weight or technology, and with that came a horsepower war. Also in the fight were Lancia, Ferrari, Jaguar, Peugeot, and Audi.

Porsche 959
1988

199^{mph}

By nature, Paul Frere is not a nervous man. Yet he was fidgeting next to me, suggesting, "John, I think the traffic is building . . ."

But I had the Porsche 959 at 280 kilometers per hour, creeping to the magic 300 (186.4 miles per hour) tick on the speedometer. The traffic wasn't *that* bad.

Frere, the world-famous race driver and automotive journalist, stirred and cleared his throat: "Ahem!" The speedo inched up to 295 . . . 300, *bingo* . . . Okay, Paul, time to slow.

These days Bentley four-door sedans top out at 195 miles per hour, but in 1989, the 959's 198.8 mile-per-hour terminal velocity was a lofty goal for a supercar. Besides, the 959 wasn't just a car aimed at a top speed—another potent drivetrain stuffed in a lightweight chassis and body—but an integrated system of systems that felt more like a performance aircraft than an automobile.

First hint at the 959 appeared as the *Gruppe B* show car at the 1983 Frankfurt Motor Show. It was Porsche's response to the FIA's Group B regulations for rally and racing cars: no minimum weight, no caps on technology, and all the horsepower you please . . . no holds barred. Build two hundred examples and you're in.

Porsche, Ferrari, and Jaguar began work on Group B cars meant for race circuits. Others went the rally route, creating all-wheel-drive, lightweight, 500-horsepower machines such as the Audi Quattro, Ford RS200, Peugeot T16, and turbocharged/supercharged Lancia Delta S4.

They created amazing cars, the Group B Rally Car website figuring an S4 got to 62 miles per hour in 2.3 seconds on gravel, and Finnish driver Henri Toivonen could have put his sixth on the grid for the 1986 Portuguese Grand Prix. But it proved a fatal recipe that year when Toivonen and his American co-driver, Sergio Cresto, were killed on the Tour de Corse. Exit Group B.

With the end of the class, Ferrari's 288 GTO and Jaguar's XJ220 were orphaned as wonderful oddities, yet Porsche's 959 became a legend.

Engineering legend Helmuth Bott headed the engineering department at Porsche as work began on the 959 early in 1983. At its heart would be all-wheel drive. It quickly became apparent the front-engine, rear-gearbox layouts of the 924, 944, and 928 were impractical for this drive layout. Ditto for development of a purpose-built midengine machine, so it was decided to adapt the traction system to the evergreen 911.

Audi made all-wheel drive lightweight and practical in high-performance road cars, its Quattro's World Rally Championship wins adding an official stamp of approval. It seemed every automaker now wanted in, and with rear-engine placement, routing power to the front wheels of the 911 was, in theory, relatively simple.

There was the usual drive to the rear wheels, with a central driveshaft carrying power to a front-wheel differential. Power had to be variable between front and rear wheels because of uneven traction possibilities. Usually a central differential did that job, but in the 959 that variation was handled by the Porsche-Steuer Kupplung—the electronically controlled PSK,

When Porsche determined that it would lose money on every 959, it decided to not certify the car for the US market. One attempt to have them imported as race cars failed. It took an act of Congress to create a new law that would allow the super Porsche (and other exotic cars) to be driven in the United States.

Porsche showed it was serious about the FIA's Group B racing class when it showed this white concept car at the 1983 Frankfurt Auto Show and, appropriately, called it the Gruppe B. Its rather futuristic Porsche styling wowed us, but what we didn't appreciate was what would lurk beneath the sleek bodywork of the production 959. Porsche's 959 wasn't just styled to look like it's going 180 when standing still; it earned that look through rigorous wind tunnel work to optimize its aerodynamics. The shape has a low drag coefficient of 0.31, and Porsche claims it has zero lift.

JAY LENO ON THE PORSCHE 959

"Seinfeld has one. They're pretty amazing, but it wasn't a car I lusted after," Leno says. Porsche didn't sell the 959 in the United States, so it was a car Leno never saw and thus never knew he might want.

Leno also explains: "They are four-wheel drive and I prefer rear-wheel drive. To me, half the fun of owning these cars is sliding 'em around. I'm not out there trying to race, I'm just out there trying to enjoy myself and have fun. [If] I go into a turn and I see a little gravel, I give it the gas and kick the ass end around and then bring it back. That's very entertaining and I enjoy that. Obviously with a 959 (and the added grip of four-wheel drive), you're going to do all those things a lot higher speed."

thirteen discs in an oil bath that could theoretically vary front/rear power distribution from 0/100 (all rear drive) to 50/50 (an even front/rear split).

The maximum split was 20 front/80 rear, the driver choosing, via a steering column lever, from four traction programs: dry or wet tarmac, snow, or off-road, each of which was indicated by a light glowing in the right-hand dash dial, adding to the sense of systems.

The suspension was a race-derived unequal-length upper and lower A-arm design with coil springs and a pair of shocks per corner. Normal ride height was 4.7 inches, though a dash-mounted switch would raise it to 5.9 or 7 inches. Above 47.9 miles per hour, the onboard computer (less powerful than a smartphone today) lowered the car to 5.9 or 4.7 inches above 99.5 miles per hour, except on the 959S (Sport) models, which had non-adjustable suspensions.

Another switch offered three shock absorber levels: soft, medium, or hard, the computer reverting to the firmest above 99.5 miles per hour.

In true Porsche form, the brakes were ventilated, cross-drilled Brembo discs (12.7 inches front, 11.9 inches rear) with four-piston aluminum calipers. In true modern form, they had ABS back when many high-performance automakers scoffed at such systems.

Ah, there's that word again—system—and here are two more on the 959 that are just getting general use today: run-flat tires and tire-pressure monitoring.

The most predictable component in the 959 was its engine, a flat-six, though it is not the 911 powerplant. Based on Porsche's all-conquering Indy/935/936/956/962 race engine, it is air-cooled with water-cooled heads containing chain-driven twin cams and four valves per cylinder. To eliminate turbo lag at low rpm, Porsche used two sequential turbos with intercoolers, the first providing boost to 4,300 rpm when the second chimed in, giving those in the 959 a nicely added oomph, and 450 horsepower at 6,500 rpm and 369 lb-ft of torque at 5,500 rpm . . . from just 2.8 liters.

Rare for a modern Porsche, the 959's engine is in plain view once the huge rear lid is raised, exhibiting a terrific scene with the vertical fan and turbo plumbing.

Somewhat unusual for its time is a six-speed manual gearbox. This wide range was considered necessary to reach the 198.8-mile-per-hour top speed and still allow for low-speed work, as it was assumed thirty years ago that four-wheel drive meant off-roading.

Arguably, the most enduring 959 attribute is the styling of its body. The 911 shape is hiding in there, but the only carryover piece is the taillight. The rest is blended into a beautiful flow of scoops, grilles, and aerodynamic detailing, some of which is still used, in forms, by Porsche today. To maintain its light weight, the 959 was based on a steel platform with the top, front fenders, rear quarters, and engine cover done in Kevlar; the doors and hood are aluminum.

Although many Porsche engines are visually lost down in the tail of the car, the 959's is proudly on display. While still a flat-six, the 2.8-liter engine has water-cooled cylinder heads, four valves per cylinder, and—to reduce response lag—two sequential turbochargers with intercoolers. Horsepower came out to 450 at 6,500 rpm and torque to 369 ft-lb at 5,500.

To get drive to the front wheels of the 911, Porsche added a driveshaft forward to an electronically controlled device at the front axle called the Porsche-Steuer Kupplung. Made up of thirteen discs in an oil bath, this PSK could vary how much power was directed to the front wheels.

More than good looking, the 959 shape has an impressive drag coefficient of 0.31 and a claimed zero lift, plus a solid stance on the road—planted to the pavement, looking as though it had all the traction you could ever need.

Interior changes were minimal, except for the tall center tunnel housing the front-wheel driveshaft and instruments, which incorporated info lights for the drive and suspension systems. Ahead are enough dials to equip an airplane, showing turbo boost to 2.5 bar, a 345-kilometer-per-hour speedometer . . . the stuff guys love. Just as they like to read about 0–60 times of 3.6 seconds and the top speed just shy of 200 miles per hour.

Racing has always been an important part of Porsche, but the 959 had limited use. A pair finished 1-2 in the rugged Paris-Dakar Rally in 1986, the same year a track version—the 961—nailed down seventh overall and first in class at Le Mans. But that was about it.

Early on, there was little doubt the 959 was headed for greatness . . . but not to the States.

Porsche planned to import the cars, but the numbers go upside down. It was costing the company a reported $500,000 to hand-build the 959 at Weissach, some $155,000 above its price tag. To minimize losses, the automaker scaled back production, slicing out the United States and keeping the original total to 226, though there was such a clamor by buyers willing to pay as much as a reported $1 million for a 959 that Porsche assembled an additional ten to fifteen in the early 1990s.

American enthusiasts wanted the 959, and the late Al Holbert, who headed Porsche racing in the United States, tried to help. He convinced the factory to build around ten 959S models as race cars, fitted with roll cages, sparse cloth interiors, 480-brake-horsepower engines, and the non-adjustable racing suspension. Holbert had to prove they were race cars, but at a test at Nazareth, Pennsylvania, the cars apparently appeared too docile, not enough like race cars for the feds, who had them shipped back to Germany.

A few 959s were imported under such legal exemptions as museum display, so famed collector Otis Chandler had one, but in the early 1990s, the situation was too fuzzy. Others, such as Ralph Lauren and Microsoft's Bill Gates and Paul Allen, not only wanted 959s, but wanted to freely drive them—which is where Bruce Canepa came in.

Representing 959 owners, Canepa had Warren Dean, a Washington, DC, lawyer, develop an exemption law and get it to Congress. Under the rule, anyone could bring in a car if it was no longer produced, there had never been more than five hundred made, and it met EPA emissions rules for the year it was made (1988 for the 959), plus an added 10 percent improvement. Department of Transportation (DOT) crash requirements would be waived, but annual mileage would be limited to 2,500.

President Bill Clinton signed the law in 1998, but it was two years before the final rule was finished, the feds being particularly fussy about that 2,500-mile proviso.

It had taken a decade, but now Canepa could import, convert, and sell 959s. The flat-six went through a development project to parallel Porsche's work, had Porsche continued with the 959. Knowing the engine and chassis had 600-brake-horsepower potential, Canepa's group reworked the electronics, replaced the KKK turbos with Garretts, and made other modifications so the 959 pumps out 610 brake horsepower with 540 lb-ft of torque, dropping the 0–60 time to just 3.2 seconds and stretching top speed to 215 miles per hour. As in the 959S versions, Canepa's cars have nonadjustable suspensions, going with modern shocks and titanium springs.

And a price tag of $575,000 with, Canepa tells us, a line of potential owners. They would join a distinguished group that already includes such men as Lauren, Chandler, Gates, Allen, and Jerry Seinfeld.

Legends with legends.

Yes, it looks like a 911, but the only piece retained from the stock car is the taillight. The remainder of the car was reworked into a beautiful flowing shape with just enough grille and scoops to make it look purposeful.

Marcello Gandini, who designed the Countach (among several famous supercars), was hired to create the shape of its successor. Called Project 132, it was taken on by Chrysler when it bought Lamborghini. Chrysler Chairman Lee Iacocca didn't like the body proposed by Gandini, and the result was a shape that combines that design with changes made to it in Chrysler's US studios.

Lamborghini Diablo 1990

205^{mph}

Ferrari was providing the perfect aural background for Lamborghini. We were at a small race course near Milan, Italy, to test the then-new 6.0-liter Lamborghini Diablo. While driving the 550-horsepower Diablo on a circuitous track, Ferrari was on another straightaway testing the software and durability of its Formula 1 car's launch system.

High revs off the line for the Ferrari, bumping up on the limiter, tires churning for grip time after time. That would inspire you.

So did the big Lamborghini. Like the Countach, there was no doubt when you drove the Diablo that it was long (175.9 inches), wide (80.3 inches minus mirrors), and heavy (3,700 pounds), and there was a lot of oomph (550 horsepower, 435 lb-ft of torque) just aft of your cranium.

A Diablo easily gets to 60 miles per hour in under four seconds, and with enough track it could trip the lights at 205 miles per hour.

It feels like it—a stormer—though with all-wheel drive you also sense security as you chase another Diablo, this one driven by a man who knows the track. You'd like to hang on a bit longer, but he's good, and your guardian angel keeps whispering, "$250,000 . . . $250,000 . . . $250,000 . . ."

Given its druthers, Lamborghini wouldn't have kept building the Countach for sixteen years, but for much of that time, it was just trying to stay alive.

Ever-changing emissions laws and the two oil shocks of the 1970s were like grenades thrown into the laps of the builders of supercars. Ferrari had become part of the Fiat empire. Maserati went to De Tomaso and was out of the exotic-car business.

In 1973, Ferruccio Lamborghini unloaded 51 percent of his car company after the first oil crisis. He dropped the remainder next year and went off to make wine. The new Swiss owners held out until 1978, when the company went bankrupt and into liquidation. A pair of wealthy French brothers, the Mimrans, bought the automaker and got it rolling successfully into the 1980s to the point where it needed additional capital.

The heart of the Diablo is its aluminum V-12 engine. With dual chain-driven camshafts on each cylinder head and fuel injection, the big engine began life in the Diablo at 5.7 liters, pumping out 485 horsepower and 428 lb-ft of torque. As in the Countach, the engine is in the car "backward" with the transmission sticking forward into the cockpit.

Providing a historical foreground for a Diablo VT is the very first Lamborghini made, the 1963 350 GTV prototype. Actually, it wasn't quite finished when presented at the 1963 Turin Auto Show, because the V-12 designed for the car by well-known engineer Giotto Bizzarrini wouldn't fit inside Franco Scaglione's beautiful bodywork.

Enter Chrysler, which purchased Lamborghini and brought with it money, expertise, and the attitude needed to convert this sometimes-sleepy automaker into an exotic-car power . . . and to design and develop the Diablo.

In 1994, Chrysler sold Lamborghini to Indonesian interests with "Tommy" Suharto at the core.

These were not the most stable of times for Lamborghini, which was pulled from the mire in 1998 when Audi AG bought it lock, stock, and assembly lines and began development of a successor to the Diablo.

But let's run the tape back to the late 1980s when Chrysler still owned the Italian automaker and it was finally time to replace the aging Countach. It was sensible that the man who created the Countach shape, making it arguably the most influential supercar of its time, should get a shot at shaping its successor. So Marcello Gandini was hired to pen a car that was codenamed the P132 but would become the Diablo.

Tempting, eh? Step in, settle down in the leather-covered seats, pull down the Diablo doors, light the fire, and take off. The ambiance is about perfect, with none of the electronic detachment of modern paddle shifters.

In 1993, Lamborghini introduced the thirtieth anniversary model of the Diablo to celebrate the company's founding. The V-12 got a boost to 525 horsepower, and the factory claimed a 0–60 time under 4.0 seconds and a top speed of 206 miles per hour. Always known for its wild color choices, Lamborghini settled on lavender metallic for the lightweight SE30 version of the Diablo. Deliveries of the 150 SE models built by Lamborghini began in 1994. Several aerodynamic changes to the SE 30th Anniversary set it apart from other Diablos. The front spoiler was deeper, with added air intakes, while new side sills directed air to oil coolers. The rear spoiler dipped at its ends, while the center was adjustable for downforce.

Things get a little hazy here. Gandini also designed the exterior of a supercar called the Cizeta Moroder V16T, and some stories have it that its design was originally meant to be a new Countach. The Cizeta was the gallant attempt by Claudio Zampolli to build an exclusive exotic machine with a 560-horsepower transverse V-16 engine. This car also fell victim to the world financial crisis in the early 1990s, but that's another book.

Bob Lutz, then president of Chrysler, once told us the company signed Gandini to do the Diablo, but what it got was a variation on the Cizeta Moroder, and the Diablo's design was completed in Detroit. Retained were Gandini's forward-tilting rear wheel arches, an alloy wheel design with holes, the forward-sloping lower window, and who knows how much else.

Regardless of exact parentage, the Diablo is a stunner. Even bigger than the Countach, it is a gorgeous design with just the right amount of strength and glamor. The design is particularly delicious when seen from above. Yes, it is even more of a job to drive than the Countach, thanks to its size, but it is arguably a more beautiful shape.

During its eleven-year lifetime (1990–2001), there were thirteen to fifteen Diablo models, depending on how you counted them, including two roadster versions. At times, the design suffered a bit from the Countach's wing-and-vented-still disease, but it wasn't as contagious or deforming with the Diablo.

As the car's tenure continued, more carbon fiber body panels were integrated into the design. With Lamborghini now well under the wing of Audi and with the 6.0-liter version in 2000, we saw the closest thing to a Diablo redesign. It was relatively minor except that most of the body, which had been in aluminum, was formed in carbon fiber, save the aluminum doors and steel structure roof.

We do know that an American, Bill Dayton, drew up the original Diablo interior. Two almost-reclining seats bracket the wide center console needed to accommodate the forward-facing transmission. Like the exterior, the car's inside is broad, elegant, and uncluttered, with the emphasis on the instrument cluster and the meaty shift linkage. When the interior was reworked for the 6.0-liter version, it had a wide arcing instrument panel and plenty of carbon fiber on display, a welcome and appropriate change.

Underneath, the Diablo is very much a continuation of the Countach. That new body shape hid a traditional steel space frame, upper and lower A-arm suspension, and disc brakes that thankfully grew over the years as horsepower increased. So did the electronic suspension content, as shocks with variable stiffness were added and then refined.

The true inner beauty of the Diablo is its V-12. Still classic Italian stuff, done up in aluminum with twin chain-driven camshafts and fuel injection, it began as a 5.7-liter with 485 horsepower at 7,000 rpm and 428 lb-ft of torque at 5,200 rpm. There were versions that really pumped it out, such as the 595-powered Jota.

Come 2000, the V-12 was stroked to 6.0 liters, taking the horsepower to 550, torque to 422 lb-ft.

Another carryover from the Countach is the reversed drivetrain that poked the five-speed transmission forward into the cockpit. This odd layout proved helpful when Lamborghini opted to adapt all-wheel drive to the big supercar.

By Lamborghini's reckoning, there was never a Diablo that wouldn't hit at least 202 miles per hour, with estimates of a few special editions edging over 210.

If the Countach was always something of a rebellious prince, the Diablo became accepted royalty among exotic cars. The rough edges had been filed off and smoothed over. Thanks to Chrysler's money and big automaker attitude, the car matured into something powerful and sophisticated and, like a few monarchs, bigger than life.

The Diablo was a solid seller for Lamborghini, which built 2,884 examples.

It wears its crown with ease.

Among the many special versions of the Diablo was the 600-horsepower SE Jota, its production run limited to fifteen cars, all finished in Electric Yellow. In addition to other body modifications, the Jota had roof-top air intakes.

Once under the umbrella of Audi, the Diablo went through one major model change while the Murcielago was being developed. The visual changes were minor, but where the body panels had previously been mainly aluminum, they were now carbon fiber with aluminum doors and steel roof.

Bugatti gave its EB110 a lavish sendoff when it debuted in 1991, parading the car from La Defense in Paris with a caravan of vintage Bugattis to the palace in Versailles. Throughout the car's short life, there was much curiosity about who was really bankrolling the production of the supercar.

Bugatti EB110
1991

212^{mph}

"You can drive at over 200 miles per hour like you're on an ordinary road," world driving champion Phil Hill said of the Bugatti EB110. "It's amazing. The steering is lovely, and the gearbox delightful. It's just an outstanding car."

Hill had just clocked 212 miles per hour in the Bugatti at Volkswagen's Ehra-Lessien track in Germany for *Road & Track*. He added, "It's the most stable car I've ever driven here."

It was never a problem to get those who drove the EB110 to praise it. Some thought clutch effort a bit heavy, and the four IHI turbos on the 3.5-liter sixty-valve (yes, sixty) V-12 didn't really get to singing until 4,500 rpm. That didn't stop *R&T* from getting an EB110 to 60 miles per hour in 4.4 seconds. As always, it was tougher to launch an all-wheel-drive car like the Bugatti off the line, but dumping the clutch at 7,500 rpm (!) worked, the car "not objecting one bit," according to Hill.

The clamor over the new Bugatti was enormous. It was the darling of the supercar set, and future Grand Prix champion Michael Schumacher bought a yellow one. The EB110 had the famed Bugatti name, a carbon-fiber chassis built by Aerospatiale (of Concorde fame), and that marvelous V-12 penned by Paolo Stanzini (of Lamborghini fame), with its 650 horsepower at 8,000 and 472 lb-ft of torque at 4,200. There was a Lamborghini door-style body drawn by Marcello Gandini (of Countach, Miura, Carabo, etc. fame) and all this was painstakingly assembled in a brand-new factory in Campogalliano, near Modena, the hometown of Ferrari, Maserati, and Lamborghini.

What went wrong?

Romano Artioli bought the name Bugatti and its famous red oval badge in 1987, promising the EB110 in 1991, the car's name and model number representing Ettore Bugatti 110 years after he was born. It made sense at the time, as the supercar world was booming. The Porsche 959 captured everyone's attention, Ferrari's F40 was fresh, and prices were flying high. But like the timing of Jaguar XJ220, the timing of the EB110 went sour.

The EB110 had the right ingredients. Marcello Gandini, who penned the Carabo, Miura, Countach, and other great exotics, was hired to do the exterior design. Paolo Stanzini, who had worked at Lamborghini, was responsible for the powerful sixty-valve V-12. And to get the power—611 horsepower in the SS version—to the ground, the car had all-wheel drive. *Richard Baron*

"You can cruise at over 200 miles per hour like you're on an ordinary road," says world driving champion Phil Hill when describing the experience of driving the Bugatti EB110. He ought to know, having just driven the car 212 miles per hour while testing it for *Road & Track*, adding, "It's just an outstanding car." *Richard Baron*

EB110s were assembled in a beautiful building in Campogalliano, on the outskirts of Modena, Italy. The huge sign behind this 1937 Bugatti Type 57S is still there, though the factory is nothing more than an empty reminder of arguably the most spectacular failure in the supercar business.

Those of us at the Bugatti's Paris debut in September 1991 were stunned by the opulence. Unveiled at La Grande Arche de la Defense, the car led a parade of vintage Bugattis to the palace in Versailles for a black-tie dinner. The apparent enormous expense of the day also raised the question of who was *really* bankrolling the venture, a curiosity that remains unanswered today.

Production began, deliveries started in December 1991, road tests rolled in, and everyone praised the car's performance. Perhaps the styling wasn't as exciting as it could be, the interior a bit pedestrian, and the car a tad heavy, but build quality was excellent and no one was complaining loudly. A second model called the SS, a somewhat lighter version with 611 horsepower, was launched in 1992. Bugatti even bought well-known automaker and engineering firm Lotus, so everyone assumed that things must be wonderful.

But then it all fell apart. The world sank into recession and supercars suffered. In 1995, the company filed for bankruptcy and the magnificent factory closed after some 125–145 EB110s were made. Ten years later, the plant remained empty, the signage a sad reminder of what never happened.

But the famed Bugatti name is not one to just lie around unused. In 1998, Volkswagen bought the rights to Bugatti and its badge, and now we have the Veyron 16.4 being built in the marque's traditional home: Molsheim, France.

Last in the line of Vectors was the M12, which was basically the Vector styling wrapped around a Lamborghini V-12 chassis. At this point, Lamborghini owned Vector, and while the M12 was not a Diablo clone, they shared many parts. One of the men who worked on the M12 exterior was Peter Stevens of McLaren F1 fame. Vector M12s feature a more conventional supercar chassis, and thanks to the Lamborghini power, they could get to 60 miles per hour in 4.8 seconds, with gearing said to be good for 190 miles per hour.

The Vector
1991

218 mph

It's easy to look back now and think we were suckers over the Vector when it was first shown, but that would be unfair.

You must know how desperate Americans were for a truly American supercar in the early 1970s. Here it was, the country that built astounding technical devices such as the SR71 Blackbird aircraft and put men on the moon, but its automobiles were technically mired in the 1950s. Italy had exotic Ferraris and Maseratis, Porsche 917s were eating up the sports car racing world, and Jaguar produced sports cars for which one would open a vein.

What did Americans have? The Chevrolet Corvette. You certainly could get a big-block 'Vette through a quarter-mile in a hurry, but the car was really a testament to Zora Duntov's ability to assemble old Chevrolet parts in some semblance of performance order, and Bill Mitchell's design department's genius at wrapping it up in a delectable package.

The automotive performance world was falling apart in the early 1970s. Laws lowering automotive emissions and increasing safety in vehicles might have been long overdue, but the effect of their ham-fisted implementation in the 1970s was to toss a huge regulatory wet blanket over performance cars. Compression ratios fell, unleaded gas became the rule, heavy safety equipment was bolted on, and engineers who were once devising ways to make engines more powerful were now just trying to make them idle successfully. Emissions laws sliced the horsepower of the small-block Corvette from 250 to 205 in 1975. No wonder Americans were discouraged.

Plus, the United States suffered two gas shocks—in 1974 and 1979—when drivers lined up for blocks to pay more for suddenly scarce gasoline. This seemed wrong in the States, home of the free, land of cheap gas.

And then, in 1978, right into this rather depressing situation, drove the Vector W2. America's savior?

We first saw a Vector at the Los Angeles Auto Expo in 1972. Designed by Gerald "Jerry" Wiegart, it had a rather amazing shape and a proposed price of just $7,500. Built in Lee Brown's body shop in Hollywood, the Vector show car had dramatic hard-edge styling with a low nose, huge glass area, and a kicked-up tail that caught our attention . . . that and its green paint, making the whole thing slightly reminiscent of the famous Alfa Romeo Carabo show car.

The color and shape earned it a place on a *Motor Trend* cover in 1973, though under the fiberglass was the tube space frame of an old midengine Dolphin sports racer and no engine. It was guessed the weight could be kept down to 2,200 pounds, and the proposed engine was either a four-cam Porsche or one of the Wankel rotary engines that were becoming the rage. (This was the period in which Mercedes-Benz and General Motors were caught in their short-term infatuation with the compact powerplant.)

That Vector project died quietly away, but in 1978 Weigert was back with a new midengine model, the W2, and it was sensational. The low (45.5 inches), wide (76 inches) brushed aluminum body with strong fender flares, a rear wing, and an imposing stance had great visual appeal. The chassis was just as intriguing, with an upper and lower A-arm front suspension and a De Dion layout at the back.

Wiegert was pushing an all-American auto-aero-tech tie in the car, making comparisons with fighter aircraft, and the interior reflected that theme with rows of switches and circuit breakers and flat-screen instrumentation.

Out back, the W2 had plenty of firepower. Turbochargers had come to the rescue of performance engine makers in this era of low horsepower, and Wiegert fitted his new car with two big ones mounted on a 6.0-liter Chevrolet V-8. Power was estimated to be around 600 horsepower and 580 lb-ft of torque. The transmission was a three-speed automatic, and top speed of around 230 miles per hour was possible, and the price tag was $125,000.

American car fans had a little faith that they might finally have a native supercar. The W2 raised interest and funding, but disappeared into a mist of rumor, speculation, and waiting.

The Vector reappeared in the early 1990s as the W8 Twin Turbo, which made it to production in Vector's factory in Wilmington, a port city west of Los Angeles. In addition to a public offering that raised $20 million, Wiegert added funds when he sued Goodyear for using the name Vector on a line of tires and won.

The styling was a little softer now, but just as exciting, still with great flared fenders, a low snout, and a better-integrated tail spoiler. The suspension remained the same, as did the powertrain, though now claiming 625 horsepower and 630 lb-ft of torque. The 3,320-pound car was still aerospace-oriented: aluminum monocoque structures off a central chrome-moly tube frame; the body finished in Kevlar, carbon fiber, and fiberglass; and the doors swinging up like a Lamborghini's. The inside-

In the W2's tail was a 6.0-liter Chevrolet V-8 with a pair of huge turbochargers. Horsepower was said to be 600, with torque at 580 lb-ft through a three-speed automatic transmission. There were claims of a 230-mile-per-hour top speed, but no one was ever allowed to test the W2 to verify this performance.

This is the original 1978 Vector W2, and atop the car is Jerry Wiegert, the man who tried to make the project work. His aim was to create the great American midengine sports car, a world beater that reflected the technology used in the great jet fighter aircraft being developed in the United States. Visually, the Vector W2 was a treat, with its brushed aluminum body. It was only 45.5 inches high, and its 76.0-inch width was emphasized by great fender flares. Its tail topped by a tall wing, the car was very imposing and guaranteed to turn heads wherever it was.

the-aircraft theme continued with toggle-covered flat panels and menus of information. All the switches and wiring were up to military spec.

The price now was just shy of $485,000, and after years of being asked, Wiegert finally allowed the car to be properly road tested. *Road & Track* got a W8 Twin Turbo to 60 miles per hour in 4.2 seconds, which was quite respectable at the time, especially on a car hampered with gearing that gave it (at least theoretically) a top speed of 218 miles per hour. *R&T* staffer Dog Kott wrote, "Utter ferocity arrives as the tach's moving-tape display skitters past 4,000 rpm, the turbos hit full stride and what feels like the hand of God presses you back into the seat. Box the shifter handle into second, the electronically controlled wastegates exhale like Darth Vader with whooping cough, and the almighty process begins anew."

Great stuff, and cars were being delivered, with a reported eighteen to twenty-two eventually built. But problems haunted the project, particularly when word got out that tennis star Andre Agassi returned his Vector because of all its problems.

Wiegert introduced a new model, the Avtech WX3, at the 1993 New York Auto Show. The chassis remained the same, but the styling was updated, still looking very exciting, particularly the back end, with the raised wing now better integrated into the body.

Funding continued to be short but got a boost in early 1993, when the company was bought by an Indonesian company, Megatech, which wanted Vector's founder, Wiegert, to step down as CEO but stay on as a designer. He resisted and was fired, but he refused to leave. Given a chance to quietly yield his spot in March, Wiegert instead used a weekend to change the locks on the factory doors, post armed guards, fire the staff, and take refuge in the buildings.

It was the stuff of daily news reports until Wiegert finally yielded and the new owners took control. The Indonesian company was owned by Tommy Suharto, son of the country's ruler, and in January 1994 he also bought Lamborghini from Chrysler. (Suharto later went to prison for the murder of a judge.)

Vector moved to Lamborghini's Jacksonville, Florida, headquarters and tried one last model, the M12, which basically wrapped Vector styling around a Lamborghini V-12-powered platform. The M12 (for midengine and twelve cylinders) wasn't a Diablo clone but was built on its own tube-frame chassis with the V-12 turned around so that transmission sprouted out the back, rather than between the front seats as in the Countach or Diablo. The De Dion rear suspension was replaced by a more conventional upper and lower A-arm design. The styling had been through a wind tunnel and then refined by two men, Peter Stevens, who did the McLaren with Gordon Murray, and Michael Santoro.

The visual result was obviously an extension of Wiegert's original theme—forward leaning, tall tail, swing-up doors—with huge holes to feed cooling ducts, but something of the early Vector's halt-you-in-your-tracks visceral appeal was missing. Ditto inside, where the car had the expected look and rich materials but not the earlier cars' tech

appeal. Gone too, of course, was any semblance to aircraft heritage or the born-in-the-USA argument.

The M12 also lost a bit of performance, with a 0–60 time of 4.8 seconds, though *Road & Track*'s Kim Reynolds suggested that number didn't reflect the M12's true, quicker acceleration potential. Handling was reasonably good, and there was talk of a 190-mile-per-hour top speed.

Vector figured it could sell 150 of its M12s per year at $184,000, built in a shop on a shuttered Florida navy base. Having seen the shop, it reminded me of one of the small English specialist companies such as TVR or Marcos. Great stuff, with sparks flying from welders' work, drivetrains being readied, virgin body panels waiting for paint, even a racing version for IMSA being readied on the side.

And that's the last we ever heard from Vector. Things weren't healthy at Lamborghini at the time, leading to its sale to Audi in 1998, and Wiegert's dream evaporated.

Or so you might think. Apparently, Wiegert isn't one to give up . . . check out www.vectorsupercars.com.

He's not letting go.

There was even a 1993 Vector Avtech Roadster. The original W2 styling from the late 1970s is still in there, but rounder and somehow even more ferocious. There is no denying the visual appeal of the Vectors, which could get any car enthusiast's heart fluttering, but years of misfired promises had made us very wary.

Wiegert's desire to give the W2 a high-tech fighter look was best seen in the interior. There was flat screen instrumentation and rows of switches. On the exterior, to go with the powerful styling and 600-horsepower drivetrain, the Vector W2 had a stout chassis. The front suspension featured upper and lower A-arms, while the rear was somewhat unusual for a midengine car with a De Dion layout.

In the early 1990s, after years of speculation, Vector was back with a car called the W8 Twin Turbo, which was tested by car magazines. *Road & Track* got one to 60 miles per hour in 4.2 seconds. The W8 was followed quickly by the Avtech WX3 and then the Avtech S/C (seen here).

While Acura is reluctant to state a coefficient of drag number, it points out that every square inch of the NSX design is meant to guide air this way or that. The Acura NSX's chief exterior designer was Michelle Christensen, part of the all-American design team that created the sports car.

Honda NSX 1991

191^{mph}

Tire sizes for the NSX are 245/355ZR-19 front and 305/30ZR-20 at the back—squat and fat to control all that power. To create the Sport Hybrid Super Handling-AWD, the front of the NSX has a pair of 36-horsepower motors, one per wheel.

Honda's first NSX put the sports car world in a tizzy when launched in 1990. Road tests so praised the new midengine machine that Luca di Montezemolo—president of Ferrari as of 1991—asked a journalist what made the Japanese car so special. He was told that Ferrari could produce a fast, fine-handling car, but "... you can't make a cigarette lighter."

His point was that performance alone was no longer enough. Even supercars had to be total automobiles with air conditioning, sound equipment and, yes, cigarette lighters as good as in any sedan.

Much of the credit for that first aluminum-chassis NSX goes to chief engineer Shigeru Uehara and designer Ken Okuyama, who penned the exterior shape of this and the Ferrari Enzo. We got the NSX from 1990 to 2005, and then it was discontinued.

We were teased with a few prototypes over several years, but they were just that until May 2016 when the first Generation II NSX rolled off the production line . . . in Ohio. Where NSX Gen I was all-Japanese, Gen II is all-American, with chief engineer Ted Klaus and designer Michelle Christensen.

That initial NSX was famous for being the first all-aluminum exotic car. Gen II goes a step further. Various aluminum materials are used for much of the inner structure, aluminum castings, stampings and extrusions, plus high-strength steel (the A-pillars).

Wrapped around this is a body composed of SMC (sheet molding compound) fenders and trunk, sheet aluminum for the doors' outer panels, and aluminum stampings for the hood and roof. That's the tech side of the body, but what we most care about is how cool it looks, hugging the road with that positive stance. Acura doesn't like to talk about the coefficient of drag—we've heard 0.30—but points out every square inch of the design is an important guide for airflow, cooling brakes or the powertrain while adding downforce. Down inside all this are the front wishbone/control arm and rear multilink suspensions, all done in aluminum. It's no surprise the brakes are big Brembos.

Seats of the NSX easily grip you and hold you in place. The infotainment system and many central controls are shared with other Acuras and Hondas.

Beautifully displayed, the twin-turbo 3.5-liter twenty-four-valve V-6 with 500 horsepower and 406 lb-ft of torque is paired with a 47-horsepower electric motor/generator between engine and transmission.

The NSX body is composed of sheet molding compound fenders and trunk, with aluminum for the doors' outer panels and the hood and roof.

We've saved the best for last. The power unit(s). Yes, plural, one half being out back, a twin-turbo 3.5-liter twenty-four-valve V-6 with 500 horsepower, 406 lb-ft of torque and a companion. That's a 47-horsepower electric motor/generator fixed between engine and nine-speed dual-clutch transmission. At the front is the Twin Motor Unit, a pair of 36-horsepower motors—one for each front wheel—packaged together.

They may be separate, but these power units work together for what Acura calls the Sport Hybrid Super Handling-AWD (All-Wheel Drive). In a sense the front and rear systems "talk" to each other, adding or reducing power to wheels for added stability front/rear or side-to-side via torque vectoring.

Belt yourself into the driver-oriented cockpit, head out on a twisty California road as we did, and the balance and your confidence level are evident from the start. The road we were on was the same one on which we drove the original NSX when new. Hard to relate directly back through all those years, but as then, the Acura was obviously as good as or better than its competition. Stable, predictable, a proper link between your hands, the tires and the road.

A worthy successor . . . starting at $156,000.

THE MODERN SUPERCAR ERA: 1992–Present

With an electronically limited top speed of 217 miles per hour, a McLaren P1 screams down the track. Perhaps this driver has made some field repairs, overridden the limiter, and is working toward 250 miles per hour.

▶

190-250^{mph}

What is it about swing-up doors that makes them so fascinating on a supercar? Gordon Murray said of the exterior shape of the F1: "It had to be a McLaren and we hadn't done one yet. I didn't want anyone to say 'I'm sorry, it's a bit like a Testarossa.'"

McLaren F1
1992

231^{mph}

Ever been to an IMAX movie, that great film system that spreads its vista of action around you? That's what it would be like to ride in a McLaren F1. With the car's unusual three-person cockpit, the driver is in the middle, and you'd be sitting to his left or right and back enough to be looking over his shoulder. In most supercars, you're next to the driver as his companion. In the McLaren, you're a voyeur to the driving process.

Peek over his shoulder, over the competition-type safety harness, and take in a whiff of the Connelly leather upholstery. Down past the Nardi steering wheel, deeply recessed away from glare, is a trio of green-lit gauges. Deeper yet, down in the footwell, are three pedals cut from aluminum and finished almost to jewelry-store standards. No rubber pedal covers, but machined faces with "F1" inscribed on each nonslip surface.

Let's say you're on the right side, so you see the tall shift lever to the driver's right and then the broad panorama out the curved windscreen. Pull down the door. The other passenger does the same. The driver twists the key, then punches the starter button. To your immediate left, the 6.0-liter BMW Motorsport V-12 kicks out. Into gear, the driver nudges the car away.

Because of your odd placement to the driver, door, wheels, engine and such, things seem slightly out of place, but there's little time for judging that, because with the throttle down the car is accelerating at 0–60 in 3.2 seconds and you're slammed back in the seat. That's when the IMAX effect kicks in, countryside flashing by on your right, trees and vehicles flitting past the windscreen, the driver working his magic to your left front, and the BMW growling away on your left.

Sensory inputs are off the scale.

I can't imagine what it's like at the F1's top speed of 231 miles per hour.

To say the McLaren F1 is special, even in the rarefied world of exotic cars, is to understate its significance. "Special" is too trivial a word. You can be special with knockout styling, an outstanding chassis, or a tricked-up engine, and the F1 has all that. But it also has something else: genius in every square inch. That's a judgment that has held up since the F1 was debuted at the Monte Carlo Sporting Club on May 8, 1992, the Thursday before the Monaco Grand Prix.

Blame the genius part on Gordon Murray. While there were others important to the F1's creation—men such as Mansour Ojjeh and Ron Dennis—the car is quite simply Murray's machine, his vision.

Other than the gray that streaks his long black hair, Murray looks little different today than when his first bit of public genius, the Brabham BT42-Ford, made its debut during the 1973 Grand Prix season. He was only twenty-six then. Season after season, Murray's cars set the standard, usually in innovation, often in successes on the GP circuit.

All this changed in 1990 when Murray left the McLaren Grand Prix design team to help establish McLaren Cars. His title was technical director, his job to create the F1, a road-going McLaren.

"Over the years, you drive many sports cars," he points out, "and every time you get in one, you note the things you do and don't like. You build up a subconscious library of what you would and wouldn't do if you ever had a chance to build a production car."

Murray begins with packaging and an amazingly low-weight target. From the outside, the F1 looks compact, and it is. As a comparison with sports cars more commonly seen, at 168.8 inches long, the F1 is 5.8 inches shorter than a Corvette Z06, though the McLaren's 107.0-inch wheelbase is 1.3 inches greater than the Chevrolet's. The McLaren is 1.0 inch narrower and 4.2 inches lower than the 'Vette. Here's the stunner: the Z06 weighs in at around 3,130 pounds, while the F1 is a mere 2,244.

Inside that compact package Murray fits three, not two, people. With a trio aboard, there's still room for 8 cubic feet of the supplied leather luggage, kept cool in storage compartments ahead of the rear wheelwells. Hold passenger count to two and it's 10 cubic feet, using a special bag that fits in the unoccupied seat. Add in underseat storage and various cubbyholes.

Typical of Murray and his creation, there's much more to passenger packaging, like that center driver's seat with its carefully determined

The F1 was almost the second-generation street machine from McLaren. Having dominated the 1967 Cam-Am season with his M6, the innovative driver/constructor Bruce McLaren considered using the M6 chassis to create a Grand Touring race car with hopes of winning Le Mans. A coupe version of the M6 was built for McLaren's evaluation as a street machine.

Built in the racing off-season of 1969–1970, the gullwing M6GT had a 5.7-liter Chevrolet V-8. McLaren was serious about the project, using the M6GT as personal transportation while it was considered as a production project. Any hope this might happen ended with Bruce McLaren's death in a Can-Am car testing accident at the English Goodwood circuit in June 1970.

From behind you can see the rear aero system of the F1 LM. The high wing adds downforce by pushing the car down to the road. Under the car is a flat bottom that allows the air to flow smoothly beneath the car, while at the back a venturi system helps draw the air out in a manner that pulls the car to the road.

JAY LENO ON THE MCLAREN

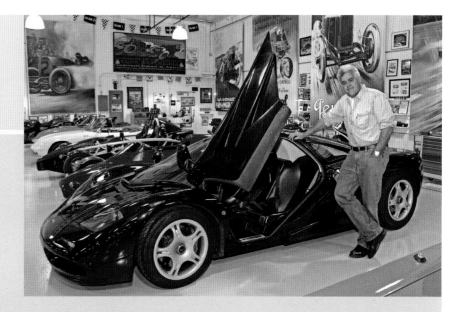

"The McLaren F1 is really sort of the greatest one of them all," Leno states, concurring with another well-known F1 owner, Ralph Lauren.

"It has 627 horsepower but weighs less than a Miata," Leno begins. In his garage, Leno parks the F1 next to another car designed by famed engineer Gordon Murray. Called the Rocket, it has a cigar-shaped body, cycle fenders, and a rear-mounted motorcycle engine, arguably the antithesis of the sleek-bodied F1, and yet Leno explains, "It's the same theory: you make it as light as possible. There's no power steering, no power brakes, no anything. The F1 is the purest driving experience of any car."

Leno also likes the fact the car is such a small, integrated package, saying, "Everything is connected, everything is structural, everything is interrelated. Plus, to me they did it the right way. They designed the chassis and engine and then built the body around it. I think the problem with the Bugatti Veyron and SLR Mercedes-Benz is they said, 'Here's what it's going to look like, make everything fit inside there.' Consequently there's a bit of a compromise. There are no compromises on the F1.

"I love the looks of the F1 because you don't know it's a brand-new car. A Countach looks period. A Ferrari Testarossa and the Miura look like an era. But the F1 could be a car from today or the '60s. Next to Ford's GT40, it's the purest design in terms of what a supercar looks like.

"From a packaging standpoint, the F1 seats three people, you've got luggage space on both sides, and yet it's smaller than a Corvette.

"And there's a great sense of theater involved with the McLaren. You open the door, it's an effort to get in, you put yourself in the center, you put in the key, you flip up the bottom . . . it's a bit like watching a pipe smoker go through all the machinations, tamping the thing, then you have the sterling silver lighter . . .

"Driving the F1 is an event.

"The downside is it's crazy explosive. There's a McLaren service center near here—that's the great thing about LA, [that] no matter what you're interested, in it's here—so I took it in for service and the guy said, 'We replaced the wiper blade.'

"I said there's no need to replace the wiper. It doesn't rain, and if it does, I don't take it out.

"He said, 'Well it's just a matter of course that we replace the blade.'

"I said, 'Okay, how much is the blade?'

"'$1,500.'

"'Don't touch the blade the next time.'

"Anytime I drive the McLaren anywhere near its limit, or my limit, I think, 'Oh, geez, if I do anything here . . . if I crack the tub, it's $300,000.' That's the problem."

That problem aside, Leno states, "It's a fascinating car and I think it's probably *the* greatest sports car of the twentieth century."

sight lines and command view. Upholstered in a leather color that contrasts with the passenger seats' black color, the bucket is carbon composite and factory shaped to fit the owner's rear end. Steering wheel and pedals are also custom installed.

Climbing into the center seat is obviously not simple, but it's well short of the gymnastic routine you'd expect. Step over, swing your butt in, and drop down. To each side are small jet-fighter-style control panels with buttons in carefully determined sizes, shapes, and places so they can easily be used without taking your eyes off the road.

On the right are heating, air conditioning, and vent controls; the right window lift; mirror adjuster; shift lever; and ignition switch and button. The left grouping has the hand brake, that side's window lift, and the CD/stereo controls. But it's not just any system. Murray recalls, "I spoke to eight different stereo companies. Most of them walked out." Why? The average weight of the quality sound system he wanted for the F1 was quoted at 36 pounds, while Murray demanded his be less than 19 pounds . . . and he got it from Kenwood.

Wrapped around this package had to be exotic bodywork, but not just a designer's flight of fancy. McLaren brought on highly talented Peter Stevens for the shape, which had to satisfy wind tunnel instrumentation as much as the human eye.

They also had a problem Murray calls "the most difficult of all: that it had to be a McLaren and we hadn't done one yet. I didn't want anybody to say, 'I'm sorry, it's a bit like a Testarossa,' or, 'It's like a Bugatti.' We couldn't afford that. It had to have a character of its own that people would identify as McLaren."

Like dihedral doors, which open up and forward, taking part of the roof and rocket panels with them to aid climbing into that central driver's seat. Or the rooftop air intake for the BMW V-12. Or the two front nostrils that take in cooling air for the aluminum radiators.

Okay, I admit it's almost pure guy stuff, but the view of the McLaren F1 LM's carbon-fiber instrument panel is like the printed instructions on a testosterone supplement. The large 9,000-rpm tachometer dominates, while to its right a 200-mile-per-hour-plus speedometer is more than just window dressing. Riding in a McLaren F1 LM is a bit like living inside a video game. With the passenger's seat slightly aft of the driver, you look over the shoulder of the gamer.

The center of the nose routes air under the car, along its flat bottom and out a tail diffuser that generates downforce. This is part of a three-element tail aero system that includes two small fans—on the same theme as Murray's 1978 Brabham BT46B F1 fan car—and a short-tail spoiler, called the Brake and Balance Foil. This strip flips up 30 degrees under hard braking to move the aerodynamic center of pressure rearward

The F1's center driver's seat makes the McLaren one of the more difficult exotics to enter. You have to step in, swing over, and settle into the seat . . . but it's well worth the effort. Drivers who raced the McLaren claim that the center spot was an advantage to them, particularly in long-distance events.

Gordon Murray wanted a large-displacement, nonturbocharged engine for the McLaren F1 and finally settled on a 6.1-liter, forty-eight-valve V-12 built by BMW Motorsports. While the head and block are aluminum, many of the small elements are magnesium. The amazingly compact powerplant kicks out 627 horsepower at 7,400 rpm and 479 lb-ft of torque between 4,000 and 7,000 rpm.

and stabilize braking from high speeds. It also raises the coefficient of drag from 0.32 to 0.40. You were warned that this isn't your average exotic car.

Murray proves this again by using carbon-fiber composite for the exterior body and chassis, the latter in combination with aluminum honeycomb. Automobile engineers of most companies in the early 1990s, when the F1 was being developed, would have chuckled when they considered the cost of even a tiny piece done in this material, but Murray had the financial freedom to dive right in.

And he did. Using McLaren International's decade of experience with carbon composites, plus computer-aided design, he created a strong occupant cell with rollover protection built into the A- and B-pillars. Immediate crush is handled by composite moldings in front and the large Inconel muffler at the back. Gasoline is carried in a centrally mounted fuel cell. The BMW test driver who walked away from a high-speed crash in the first F1 prototype can happily attest to the car's safety.

Although the shape needed for the racing McLaren GTR was fundamentally the same as the street machine, there were detail changes. Most prominent in this view was the addition of a tall rear spoiler for downforce, but you can also see vents atop the front fenders and altered aerodynamic detailing around the front wheels and along the bottom on the sills.

This carbon composite structure also gives Murray a very stiff chassis for the F1's suspension, which was developed with engineer Steve Randall. At first glance, it is somewhat conventional but attached in an unconventional manner. Up front are unequal-length A-arms attached through plain bearings to a subframe. This frame is then mounted to the carbon tub via four elastometric bushings that work at different angles—a McLaren-patented system Murray calls Ground Plane Shear Centre. While the arms are mounted to the subframe to give the geometry control needed for precise handling, the bushings damp out noise and vibration. The combined Bilstein spring/shocks are mounted horizontally and operate through rocker arms off the upper A-arms.

Unwilling to accept even the smallest deflection caused by a softly mounted steering rack, it is an integral part of the cast front bulkhead. Incidentally, the rack-and-pinion steering has no power assist.

Basics of the rear suspension are again upper and lower A-arms, here mounted to the transaxle and engine.

In concert with the Brake and Balance Foil system mentioned earlier are the huge drilled Brembo brakes, 13 inches front, 12 inches rear. McClaren wanted carbon brakes but at the time couldn't make them practical for street use. They did add a unique touch with intelligent brake cooling, electronics deciding when to open brake cooling ducts, which would otherwise affect aero, at high speeds.

McLaren's traditional GP tire supplier at the time, Goodyear, provided the F1's special skins, and OZ Racing created the magnesium alloy wheels.

Early chassis development work was done with a pair of steel-frame mules royally named Edward and Albert. While the carbon composite monocoque was still being engineered, these two prototypes gave McLaren several years of test service. Albert had a big-block Chevrolet V-8 for its horsepower, while Edward was fitted with what would become the F1 production engine, a BMW V-12.

Given Honda's Grand Prix ties with McLaren until 1993, it would seem natural for the Japanese company to supply the F1's engine. There was considerable discussion with Honda, but Murray wanted a normally aspirated powerplant and, as the projected displacement grow out of the 3.5- to 4.0-liter range, Honda's interest waned.

What most interested the engineer was a V-12 and plenty of horsepower, which BMW was able to offer in a 60-degree 6.1-liter forty-eight-valve V-12 through its Motorsport division. This engine had little to do with V-12s found in 750i and 850i BMWs of the era and was purpose-built for McLaren.

The head and block are aluminum, while many other castings—oil pump, cam covers, and the housing for the variable valve timing—are in magnesium alloy. It's a surprisingly compact powerplant, its height trimmed with dry-sump oiling. Topped by its carbon composite airbox, which shows its fabric skin pattern, with cam covers that declare "McLaren" and "BMW MPower," and with almost artistically twisted exhaust headers, the V-12 is a thing of mechanical beauty. And power, with 627 brake horsepower at 7,400 rpm and 479 lb-ft of torque between 4,000 and 7,000 rpm.

Remember the F1's long wheelbase measurement for its overall length? They get that minimal rear overhang thanks to a shockingly small transverse six-speed gearbox, which doesn't have to hang out the back. Between the engine and gearbox is a tiny aluminum flywheel and (another carryover from GP racing) a three-plate carbon-on-carbon clutch, a combination that provides minimal mass and inertia. And if you've ever fumbled your way through the muddled gear change in a midengine car, even some expensive exotics, you'll appreciate Murray designing a system that can get you down to racing gear-change times.

So is that all you got when you plunked down the $1 million for your McLaren F1? Nope. There are also Facom tools—a small titanium set for the car, a floor unit for your garage, and a car cover. When the McLaren F1 went on sale in 1992, it quickly snapped up all the acclaim expected of it . . . and all the sales, with customers more than happy to stand in line to pay $1 million for one of the sixty-four street F1s built.

But the story doesn't end there.

It was never intended that the F1 should be raced, but given its performance potential, several owners went to McLaren's Ron Dennis requesting a racing version. At first he refused, but when the owners then suggested they would do it on their own, he relented and the F1 GTR was created for 1995.

The main objective of the exercise was the BPR GT Championship and, more important, the 24 Hours of Le Mans. Regulations allowed some changes, but surprisingly few were made. Ride height was lowered, the suspension was firmed, and brakes were upgraded to carbon discs inside larger wheels. An external rear wing was added for more downforce and the bodywork revised slightly for better high-speed aerodynamics and cooling.

Horsepower was reduced. Rules required an air restrictor that cut the BMW V-12 to 600 horsepower, so by the time you add the drag of the wing, the race car was slower than the street version.

There were the expected changes inside: the amenities gone, a roll cage added, race instrumentation, and such.

F1s won the BRP GT Championship in 1995, 1996, and 1997, with the car going through more development each year as the rules were revised to allow more deviation from stock. By 1997, there was a long-tail F1 that weighed just over 2,000 pounds, but as the rules got broader still, Porsche and Mercedes-Benz jumped in with cars that had little to do with production machines, and the series suffered.

Along the way, however, McLarens were amazing at Le Mans, where just finishing is a victory. In 1995, the cars finished first, third, fourth, and fifth. The next year they were fourth, fifth, sixth, eighth, and ninth, with a second and third in 1997 and a fourth in 1998.

Pierre-Henri Raphanel, now the *pilote official* for Bugatti, shared driving duties in the second-place car in 1997, when McLaren also won the GT class. Asked what was so special about the F1 GTR, he points out that the car's systems—engine, brakes, suspension, steering—were in good harmony, not as in some race cars where the engine power, handling, brakes, or gearbox might dominate.

He then points out that the greatest advantage was the central driving position. In other race cars, the driver has to allow for more race car space on the right or left side (depending on whether the car is left- or right-hand drive); that wasn't an issue with the McLaren. He says it was a great advantage, particularly in long-distance races.

McLaren built a series of five special F1s in 1995 to commemorate its Le Mans victory. Called F1 LM, they were basically GTRs made streetable, still with the fixed rear wing and straight-cut gears, but minus the air intake restrictors, so horsepower jumped to 680. Even rarer were the three F1 long-tail street machines built to homologate the revised bodywork for 1997, fitted with bodywork that extended both the nose and tail for aerodynamic efficiency.

One of the F1 LM owners is one of the world's great McLaren fans in addition to being one of the world's great fashion designers. Ralph Lauren owns three F1s: two street machines and one of the Le Mans replicas

 McLaren had no plans to race the F1 and at first resisted pleas by owners to develop the supercar for competition. But when a trio of owners told Ron Dennis, who managed McLaren, that they would be racing their cars regardless, the company relented and developed the GTR racing version.

built to celebrate the 1995 win. He has an extensive collection of great automobiles, but Lauren explains, "Once you drive the McLaren F1, it's over. It's like no other car I've ever driven . . . it's *Star Wars*. A hovercraft. I feel like I'm not touching the ground, like the road is down there and I'm up here. It's an experience I've never had before."

Which probably wraps up the opinion of anyone who has ever driven or ridden in a McLaren F1.

Like its successor, the Enzo, Ferrari's F50 has numerous ties to the company's Formula 1 car. Unlike the Enzo, which was shaped to reflect the feeling of those open-wheel cars, the F50's Pininfarina-penned exterior looks more like a Group C sports racer.

Ferrari F50
1996

202^{mph}

Ferrari's 1961 world driving champion, Phil Hill, was driving while I was happily belted into the passenger's seat as we set out to lap Ferrari's Fiorano test track for a story for *Road & Track*.

Unlike some supercars, this was a somewhat raucous ride. For one thing, we were out in the open, and the nature of the F50 was different from many of the others. This one was very much a race car for the street, one you could feel and hear all around you in the sounds that swirled forward from the V-12 and the vibrations and feedback that drummed up from the road. It accelerated in a slightly more brutal manner.

When Phil was hard on the brakes, I was hard forward in my belts as I rode down with the car. Then on to Fiorano's straight, aimed under a bridge and past the pits, where we wouldn't match the car's top speed of 202 miles per hour but were well into three figures. The air over the car was at near-hurricane strength.

When you drive the F50, it tousles your hair, tingles your soul, and gives you a profound sense of what it would be like to ride in a great sports racing or possibly a two-seat Grand Prix car.

Having a Formula 1 machine for the street is the stuff of many men's dreams. And since Ferrari has been fulfilling men's driving dreams for years, it only seems appropriate that the Italian company would base its 1996 supercar, the F50, on Grand Prix thinking.

That number, fifty, not only coincided with the automaker's birthday, but also happened to be the natural numerical successor to the famous F40. And, like any proper successor, it was a significant step forward.

We begin with the part of the F50 that can't be like a GP car—the configuration and body. Pininfarina, the traditional designer of Ferrari exteriors, also did the two-seater F50 and needed to make it both exciting to look at and aerodynamically efficient. While one wouldn't call the F50 a beautiful car in the manner of Ferraris like the 250 GTO, Lusso, and Daytona, it is undeniably exciting . . . it stirs your soul. The basic shape—done in carbon fiber, Kevlar, and Nomex honeycomb—tells you it's a midengine design and appeals to your heart. And to your head, because you know Pininfarina wouldn't let this car out of the studio without proper wind tunnel development work.

Carbon fiber, Kevlar, and Nomex were used for the F50 body. The hood was formed with heat-relieving vents to help cool the front radiator. Airflow is directed by the shape of the nose, while at the back, the tall spoiler adds downforce. Coefficient of drag of the F50 comes in at 0.372.

Ferrari opted for a naturally aspirated 65-degree V-12 for the F50, an engine that shares fundamentals with the company's successful 333SP sports racing car. The sixty-valve engine comes rated at 513 horsepower at 8,000 rpm and 347 lb-ft of torque at 6,500.

As a result, the bonnet of the F50 has been shaped both to relieve heat from the front-mouthed radiators and provide front downforce. The leading edge of the front bumper separates the air that flows over the car from the flow that is sent underneath and out the channels at the back of the car's underside, providing more "stiction" to the road.

At the back, the tall spoiler adds a proportional amount of downforce to match the front's push-down on the road. Coefficient of drag is 0.372. In a clever concession to complaints that top-level Ferraris should be open cars, the F50 can be used as an open roadster Barchetta or with a hardtop in place, creating the Berlinetta version.

Whether the wind is in your hair or you are cozy under the hardtop, the interior is the same, mixing the new and the old at Ferrari by emphasizing the use of both leather and visible carbon fiber. The seats are grippy and come in two sizes, normal and larger. Based on shells of composite plastic, the seats are trimmed in leather and a breathable cloth. Naturally, the driver's seat moves for adjustment, but so do the pedals, all the better to suit the pilot's size.

At the driver's right is the gearshift, with a carbon-fiber knob and the classic metal shift gate. Again, following F1 practice, but breaking with

Ferrari road car tradition, the instrument panel is an electronic LCD display lit by electroluminescent bulbs. Naturally, the main dials are the tachometer and speedometer, but also on the display are oil pressure and temperature, coolant temperature, and, of course, fuel level. Would-be racers like the FIA-standard roll bars and four-point seat harnesses, but the F50 also has such amenities as climate control and leather pouches for odds and ends, plus a courtesy light . . . which is not F1 practice.

Back in the Grand Prix frame of mind, the F50's chassis is based on a carbon-fiber tub that weighs in at only 224 pounds but is quite stiff in tension. Like a race car, the Ferrari's fuel cell (not a tank, mind you, but a proper rubber compound cell) is located behind the driver-passenger compartment ahead of the engine and centralized away from as many accident impacts as possible.

Look at the front of the bare F50 tub and you will see steel plates where the suspension is attached. In back, the suspension is bolted to a piece that fits between the engine and gearbox, which are rigidly mounted to the tub, housing an oil tank, another F1 holdover. Both suspension designs are based on, in race car practice, upper and lower A-arms with push-rod-controlled shock absorbers. Those shocks, incidentally, are electronically controlled to vary the damping with the driving and road condition.

Other chassis components include a rack-and-pinion steering system designed by TRW and cast in aluminum alloy. Going to its racing supplier for the brakes, Ferrari used Brembos with aluminum calipers gripping cast-iron discs that are cross-drilled for better cooling. With diameters of 14 inches front and 13.2 inches rear, the disc brakes are large enough to haul the F50 down from all conditions without needing any sort of power boost. Because of this and the nature of the Ferrari supercar, anti-lock is not part of the F50 braking system.

Other important elements of the F50 design are, of course, the wheels and tires. Those Speedline wheels do double duty, their star-shaped

design adding to the aggressive visual range of the Ferrari, their function aiding the handling. Fitted to those wheels are another (at that time) Ferrari-F1 tie—Goodyear tires. The big "skins," measuring 245/35ZR-18 front and 355/30ZR-18 rear, were specially developed for the F50 and named after Ferrari's private test track, Fiorano.

While Ferrari likes to make a great deal about the ties between the F50 and Formula 1, the fact is the supercar also shares a great deal with Ferrari's 333SP world championship sports car. That's where you find much of the heritage of the F50's 4.7-liter V-12 engine. The heart of the engine is the block, its two banks of cylinders separated by 65 degrees. There's a seven-main bearing crankshaft with titanium rods and aluminum pistons. The oiling system is by dry sump.

Ferrari seriously considered racing the F50 and built three competition versions. Weighing 2,000 pounds with a 750-horsepower V-12 matched to a six-speed sequential gearbox, the car was very quick around Ferrari's Fiorano test track. The automaker decided to scrap the program to concentrate on its Formula 1 program, a decision that resulted in multiple world championships.

Quite a view, right? That's World Driving Champion Phil Hill behind the wheel and Editor-at-Large Peter Egan riding as they do hot laps at Ferrari's Fiorano test track. Thanks to an easily removable hardtop, the F50 can run as an open Barchetta or closed Berlinetta. Ferrari named this car the F50 to honor the famous Italian specialist automaker's fiftieth anniversary. The F50 presents quite a contrast to the front-engine, 2.0-liter cars that marked the beginning of Ferrari's history.

Atop each cylinder bank, two camshafts on each head open five valves per cylinder. By using five valves, they can be made smaller and more able to take high revs—over 10,000 rpm—without floating. Feeding fuel and air past those valves is a Bosch Motronic 2.7 engine management system looking after the electronic fuel injection and static ignition.

Another interesting feature of the F50 engine is the exhaust system, which uses two different exhaust system lengths. One length gives the sort of all-around low-end torque that is expected from even high-rev modern supercar engines. The second length is utilized automatically by the electronics and lessens back pressure in the exhaust system for more power at top speed and under full load.

All this is bolted together with an 11.3:1 compression ratio, giving the F50's V-12 engine 513 horsepower at 8,000 rpm. Torque is 347 lb-ft at 6,500 rpm. Getting all this to the wheels is a six-speed manual gearbox. Housed in a magnesium alloy case, the gearbox has an oil cooler and a limited-slip differential but no traction control.

Unlike the F40, which needed three years for US certification, the first F50s were shipped to the States, in part to dive in before a new stricter emissions law.

After that, the factory worked away until 349 of the F50s had been built. And that was it. Critics felt Ferrari milked the F40 by building too many— more than 1,300. Ferrari didn't with the F50 but hinted at another super Ferrari in the near future . . . which turned out to be the Enzo.

Ferrari also considered racing the F50 and built a trio of competition F50 GTs. Weight was cut to under 2,000 pounds and horsepower upped to 750 with a six-speed sequential gearbox. Some reports had the F50 GT lapping Fiorano faster than the 333SP sports racer, but Ferrari decided to concentrate on Formula 1 and, after building only three F50 GTs, sold the cars and bagged the program.

Sad, because on those laps with Phil Hill around Fiorano, there was the sense that the F50 was more than just a street machine—that it was a thoroughbred just waiting to be let loose. And it would be out of the gate in a hurry, as Road & Track had timed the F50 to 60 miles per hour in just 3.6 seconds earlier in the day. Ferrari said the car was good for 202 miles per hour, though our expert witness figured the gearing for that acceleration might yield only 190. Oh, darn . . . only 190 miles per hour.

Just before the finish of the day, Phil took R&T Editor-at-Large Peter Egan for a ride in the F50. For the first few laps, we led them around for photos, me in the trunk of another car. They drove up close for a car-to-car shot, the F50 just inches away as we twisted around Fiorano.

What a sight . . . I could have sold tickets.

The late Paul Frere, renowned automotive journalist and race driver, with a Pagani Zonda on the Futa Pass, which was at one time part of Italy's famous Mille Miglia open-road race circuit.

Pagani Zonda 1999

214^{mph}

You see it before you hear it—the silver Pagani Zonda slipping through the Italian countryside at speed. Driven by noted automotive journalist and race driver Paul Frere, the rare supercar weaves through a downhill set of esses, quite gracefully for its speed, the velocity not as apparent as you think. It flies by you like an elegant burst of energy and disappears.

Le Mans winner Frere figures the Zonda "provides excellent comfort by supercar standards, together with superb handling."

There is also a nasty side to the car, with 0–60 miles per hour in 3.5–4.0 seconds, depending on which version of the Zonda you have, but top velocity is on the uphill side of 200 miles per hour.

For most drivers, a dream car is one they wish to own, while Horace Pagani's dream was to build one. An Argentinean expatriate living in Italy, Pagani learned the fiberglass then carbon fiber trades, became an expert, and had the time and resources to design his car. And we mean the entire automobile, from its exterior shape to the midengine chassis to manufacturing in a factory he laid out.

His major outside supplier is Mercedes-Benz, which provides its latest AMG V-12, an excellent choice for its high-horsepower, certified emissions controls, and quality. In his plant near Modena—the Italian supercar Garden of Eden—Pagani builds the entire car, which is basically carbon fiber. There are aluminum subframes front and rear to carry the suspension and drivetrain, but the rest is of the lightweight composite. Particularly slick looking are the few cars he has finished with unpainted carbon fiber patterns as the exposed finish. That exterior shape is a bit quirky from some angles, but very cool looking and a celebration of details, particularly at the back end.

You can order your Zonda ($500,000–$600,000) as either a coupe or roadster. Horace Pagani says the chassis reinforcement needed to a decapitate the Zonda costs a surprisingly low 44 pounds and the 11-pound carbon fiber top not only can be installed or removed in minutes but also stays in place when traveling over 200 miles per hour. That's no simple trick.

Inside, the Pagani is again a bit quirky, as the air vents look a bit like factory ducts; it has huge toggle switches and a festival of carbon fiber, but it's fun. There are plenty of details to catch your eye and seats that sit as good as they look.

We first found the Zonda C12 at the 1999 Geneva Motor Show, and since then have seen five more iterations that are best described as "the same, only different." The body style remains unchanged, as does the chassis with its upper and lower aluminum A-arm suspensions, rack-and-pinion steering, and large disc brakes. Also unchanged is the light weight, because Horace Pagani takes full advantage of carbon fiber's properties to finish his cars in the 2,800–2,900-pound range and keep them impressively rigid.

What varies is the version of the Mercedes-Benz AMG V-12 out back, though you could plan on 7.0–7.3 liters, 550–600 horsepower, and an equivalent amount of torque in lb-ft form the aluminum twin-cam engine. Pagani opts for a six-speed manual transmission.

The first Zonda, the C12, was presented at the 1999 Geneva Auto Show. Given the nature of the exotic car business, especially for the very small, specialist firms, it's impressive that Pagani is still in existence.

Thanks to the clever use of carbon fiber, Pagani is able to keep the weight of his cars to less than 3,000 pounds. As these are personalized one-off cars, owners can virtually custom-order their Zondas with various choices of engines, colors, and such.

The most recent edition is the first Pagani truly meant for the United States: the C12F, the last letter in honor of Pagani's fellow Argentinean, the great race driver Juan Manuel Fangio. You have the choice of a 602-horsepower model or the Clubsport with its 650 horsepower and 0–60 potential of 3.5 seconds. And beyond that? Pagani claims 214 miles per hour.

Why would an exotic-car buyer be interested in the Pagani? It certainly is unique, with an estimated sixty built between 1999 and the beginning of 2006. Horace Pagani is a very likeable guy, and there is a certain long-term honesty to his car. Every expert who drives it comes away impressed and, unlike some very-limited-production automakers, there is no sense of scam or scandal about the place. No salacious rumors about funding or the future.

Best of all, at speed the Pagani Zonda just flies by you like an elegant burst of energy and disappears.

Horace Pagani grew up in Argentina but moved to Italy to fulfill his dream of working in the car business. After becoming an expert in fiberglass and carbon-fiber work, he was able to create his own exotic car, the Zonda, which he builds near Modena, home of Ferrari and Maserati. He drew the shape for the Zonda body, which is striking from any angle. But Pagani also designed what's underneath that carbon-fiber exterior—at the car's core is a carbon-fiber center structure with aluminum subframes front and back to accommodate the upper and lower A-arm suspensions and the V-12 drivetrain.

Interiors in the Zondas can be quite spectacular. There is plenty of carbon fiber showing in the roadster, nicely contrasting with the leather seats and brushed aluminum center dash panel.

Phil Frank, working with Steve Saleen, gets the nod for the S7 body, which is as stunning as anything out of Italy. The low nose, the dramatic air intakes along the sides, the arcing vents, the roof-top air intake, and that sweeping tail are all done in lightweight carbon fiber. Flip-up doors and beautiful exterior and interior finish aside, under that body the Saleen S7 is race bred. The upper and lower A-arm suspension is adjustable, and many of the parts have the machined beauty of serious race cars.

Saleen S7
2001

235^{mph}

I was at the 12 Hours of Sebring in Florida at sunset, down at the famous hairpin. Now was probably the best time to see this famous endurance event, the race cars showing some wear, a light coat of grit on their once-shiny finishes as they braked hard for this tight corner. The sound of their open exhausts rumbled around me as they dived for the hairpin and, in the dulling twilight, their disc brakes glowed bright orange.

Audis, Aston Martins, Ferraris, Porsches by the score, and a car that didn't come from any foreign land but the California coastal city of Irvine: Saleen's S7.

Corvettes and Saleens have been upholding the honor of the US racing community for years. The C6R 'Vettes are beautifully developed, highly specialized race versions of America's favorite production sports car. Saleen's race machines, on the other hand, are not that far removed from the S7 you can drive on the street.

Saleens are winners, having taken their class in the rough 12 Hours of Sebring, competed with honor at Le Mans, and captured driver's and manufacturer's GT championships in the United States, England, and Spain.

In fact, if you ever wanted to know what it's like to drive a race car on the streets, finagle your way into a Saleen S7. This is the real thing, now with a 750-horsepower twin-turbo V-8 that gobbles up 0–60 in 2.8 seconds and the quarter-mile in 10.7 seconds at 136 miles per hour. *Road & Track* discovered that from a standing start an S7 covers a mile in just 23.4 seconds, crossing the line at 205.7 miles per hour. Steve Saleen figures that if you don't lift and have enough road you'll redline at 235 miles per hour.

When you get into the S7 via Lamborghini-style flip-up doors, if you have broad feet like me, you may have to remove your shoes to fit into the footwell. The procedure seems daunting but isn't. Once in, you are snug, but not squeezed into what is a very personal two-seater. The driver's seat is fixed, but the pedals can be manually reset fore-aft and the steering wheel tilt is adjustable.

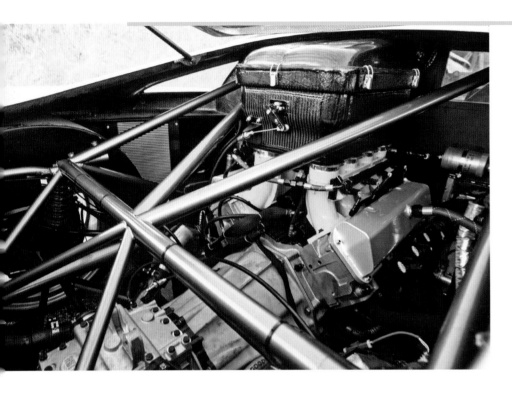

The route used to familiarize S7 owners with their new car is a course of lightly traveled roads and motorways, which quickly tempts the speedometer into three figures. As you pass between 90 and 100 miles per hour, you can feel the downforce a bit as the S7 settles in and steadies up still more. At 125 miles per hour, the car is rock-solid and the temptation to keep accelerating is soooooo great.

Despite being race car-based, the S7 is easy to maneuver in traffic, though you feel a bit down-there on the pavement, looking up at not just trucks but even Toyota Corollas. Outward visibility to the sides and front is reasonably good, vision to the rear handled by a small video camera and dashboard-mounted screen.

The ride is what you'd expect from a race car-turned-road machine: very firm, though the reality of what you're driving seems to make you more understanding and amenable to the snug-down nature of the car. Sounds of the car hitting road imperfections bring you back down to earth.

Included on the drive is a twisty, up-down, left-right forested two-laner on which the S7 proves quite docile, feeling like it weighs about half its real-world ton-and-a-half. It seems that the faster you go, the narrower

The close-in proximity of your surroundings shouts race car, but the cockpit image says luxury, with fine leather, a proper center console, and watch-face gauges, all nicely finished and put together. The presence of amenities such as air conditioning, power windows, and a sound system might lull you into luxury, but at the push of a button, the V-8 startles awake behind you as a friendly-but-growling presence.

It's a heavy-clutch—though not dauntingly so—and requires a balance between revs and take-up that at first has you feeling you've spun it too much . . . or stalled the engine. Once moving, the shift linkage is positive and you can feel there's enormous power just aft.

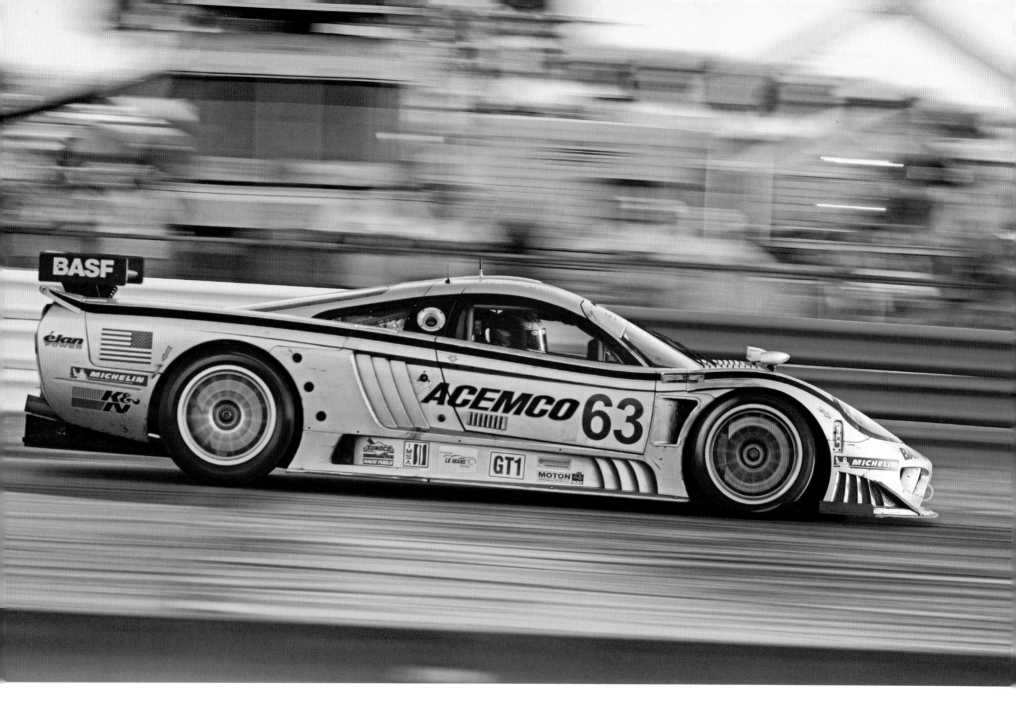

Okay, here are the physical facts: Curb Weight: 2,950 pounds. Weight distribution: 40 percent front/60 percent rear. Wheelbase: 106.30 inches. Track: 68.82 inches front/ 67.32 inches rear. Width: 78.35 inches. Length: 187.95 inches. Height: 40.98 inches. Best of all, the W8 Twin Turbo gets those dimensions to 60 miles per hour in 2.8 seconds.

Some car fans have trouble taking Saleen seriously, because the car doesn't come from Italy or Germany, which is about as silly and short-sighted as not taking a Corvette Z06 seriously. The design, the detailing, and the execution of the California-built cars are up to that of the imports. And like Ruf in Germany, Saleen has achieved full manufacturer status, doing the requisite testing for emissions and safety. Saleen has long had a tight relationship with the Ford Motor Company and builds the Ford GT for the Dearborn automaker in a factory near Detroit. You can drool more over the S7's details by checking in at www.saleen.com.

the car becomes. Turn-in with the light, quick electric-hydro assist steering is a breeze; just point and you are there.

You could drive on forever.

For the past twenty-five years, Steve Saleen has been successfully racing and building Ford-based cars.

Like Louis Ruf in Germany, Saleen isn't a tuner but a manufacturer, right down to his own vehicle identification number (VIN) plates in each car. Most of the employees in Saleen's 120,000-square-foot factory in Irvine are making up to four hundred changes to Mustangs, converting as many as one thousand each year into Saleen versions.

Because these modifications are so extensive, Saleen had to become a manufacturer and do both crash testing and emissions certification

of his cars. He does this with Ford's blessing, selling his cars through franchised Ford dealers, complete with warranty.

Saleen is so tight with Ford that his factory was the skunk works for the Ford GT project, and he is manufacturing the midengine exotics for Ford in a facility just outside Detroit.

The S7 is not, however, the Ford GT in different clothing, but its own animal, constructed race car-fashion in what looks like a competition shop, where each car's basic structure is welded up of chrome-moly tubing reinforced by aluminum honeycomb panels. The results are so sturdy that when you ask Saleen about meeting government crash standards, he grins and quips that it's not difficult with a structure meant to protect a race driver who might leave Le Mans' Mulsanne Straight at more than 200 miles per hour.

Suspension layout is as expected of a race-bred car: upper and lower A-arms and dampers with coil-over springs; but what impresses is the quality of workmanship—the CNC cut-from-a-billet aluminum front hubs and the fabricated suspension arms. And like any proper race car, the suspension is fully adjustable.

Vented Brembo non-ABS four-piston caliper brakes, measuring 15 inches front/14 inches rear, lurk inside wheels with squat Michelin Pilot Sport tires mounted to alloy wheels with race car single center nut mounting.

Ford parts are a few of the beginnings for the aluminum sixteen-valve V-8, but any comparison quickly fades when you see how the big engine is built. Saleen engineers have kept some Ford dimensions, such as the bore centers, so they can use the best aftermarket racing gaskets, but the aluminum block and heads, plus stainless-steel valves and such, are custom-made for Saleen, complete with special touches such as beryllium exhaust valve seats. The dry-sump engine sits very low in the car, making it about the toughest to photograph in my experience, and it is fed fresh air from a scoop at the front edge of the roof.

The S7 I drove was the original with a mere 550 horsepower, but as of 2005 Saleen began fitting the big motor with a pair of turbochargers, sending horsepower sky high, straight to 750 at 6,300 rpm with torque an equally impressive 700 lb-ft at 4,800. And this in a car that weighs 2,950 pounds. No wonder it gets to 60 in 2.8 seconds.

Around all this is a very dramatic carbon-fiber body, shaped by Phil Frank and Steve Saleen and meant to yield the combination of sleekness and downforce needed in a high-speed GT. Aero effectiveness aside, it's a thrilling form that looks like it should cost the $550,000 Saleen charges. The shape has a high louver and scoop count, for one purpose: to aim airflow in the proper direction, from the gaping nose to the airflow exhaust diffuser under the rear spoiler.

As I said, it's the real thing and makes you wonder what it would be like to make the right-turn hairpin at Sebring in a Saleen S7 and just keep going . . .

When *Road & Track* tested the Enzo, it generated a very impressive 1.01 g on the skid pad and snaked through the 700-foot slalom at 73.0 miles per hour. The Enzo has a surprisingly comfortable ride for a car with such high handling potential.

Ferrari Enzo 2002

218^{mph}

This is how you drive a Ferrari Enzo: Yes, the car is wide, but so is the door, and when it swings up and forward, the structural sill you step over to get in is surprisingly narrow. Settle in, belt up, step on the brake. Turn the key to the right. Pull both paddles that electro-hydraulically shift the six-speed transmission back to get to neutral. Push the start button. Tug back on the right paddle for first, off the brake, and drive away.

As the revs wrap up on the redline at eight grand, pop back on the right-hand lever—don't even lift off the gas—and you're in second, though you'll be back at 8,000 in two quick breaths, with the V-12 singing its heart out behind you.

You pop into third gear at 75 miles per hour, and sooner than you realize, you are at criminal-offense speeds and back off. It's sooo easy, and there are more gears between you and 281 miles per hour.

This does not mean that you can go to Ferrari's Italian test track, Fiorano, and start challenging Grand Prix legend Michael Schumacher's lap times. That's another matter altogether—like needing serious driving talent—but at least you can get up a head of steam in a heartbeat and even sound like you know what you're doing on downshifts without any of the old impediments.

In fact, the Enzo is so easy to drive your mother could do it . . . if she had the $1 million to buy one.

How I learned all this about the Enzo is another story. When the supercar was launched in 2002, Ferrari had no test examples, so *Road & Track* had to find an accommodating owner. Enter Richard Losee of Provo, Utah, an old friend of the magazine and an Enzo owner. He offered the car for testing as soon as he got it, but it needed pre-track miles, so would we come to Provo and help him drive it to Newport Beach, California—*R&T*'s home—to put 1,500 miles on it?

You bet. Four of us, world driving champion Phil Hill, Design Director Richard Baron, Road Test Editor Patrick Hong, and lucky me, went to Provo so we could drive the Enzo back. And it was "we" because the ever-generous Losee wanted all of us to drive the car.

Ain't life grand?

In Provo, we found it a bit shocking to see an Enzo on the street for the first time, even for those of us who love the design. It would be like seeing a Martian, though an attractive one, at the grocery store. It *really* stands out in a crowded road.

You can read about the Enzo's creation in the sidebar story with its designer, Ken Okuyama, but there is also the aerodynamics story.

If you stood back, climbed a 6-foot ladder, and looked down on the Enzo, you could easily see that the shape was like a dramatic shrink-wrap around an F1 machine. The design, however, was meant to do more than just attract the eye.

Ferrari and Pininfarina wanted a 200-mile-per-hour-plus exotic car without the large external wings seen on earlier supercars, so the Enzo went through extensive wind tunnel aero work. You would find intakes and outlets at important points on the body, all meant to channel airflow precisely and contribute to aerodynamics and the cooling required for a 650-horsepower high-performance engine to propel an Enzo quickly to more than 200 miles per hour.

The nose design contains a hint of a Formula 1 car's wing, while inside that low nose are hidden flaps that open and close to aid cooling and downforce. The small rear wing rises at 37 miles per hour to aid and balance downforce but retracts in steps as velocity increases to trim the car's shape for ultra-high speed. Ferrari pointed out that the downforce increased from 758 pounds at 124 miles per hour to its max at 1,709 at 185 but lowered to 1,209 pounds to cut drag and get the Enzo to its top speed of 218 miles per hour.

An important part of the downforce package is hidden under the car: a flat bottom with a venturi under the rear bumper that helps suck the car down to the road. I followed the Enzo on a very dusty road and had a dramatic demonstration of how the system vacuums the air from beneath the car.

Up on that ladder you'd also be looking at a lot of carbon fiber, because all the body panels and the inner monocoque tub that is the structural heart of the Enzo are made of carbon fiber with appropriate aluminum honeycomb panels. This F1-like tub was computer-designed for low weight and high strength. It weighs an amazingly low 202 pounds. This structure design is also critical to safety, as an Enzo owner found out in Malibu, California when his car broke in half against a power pole at 125 miles per hour (or more) and he walked away with just a bloody lip.

Aluminum subframes were included to attach the driveline and suspension to the tub.

Computers also aided in the design of the upper and lower A-arm suspensions with their pushrod-activated coil springs/shocks, which are located inboard horizontally. To stop its 200-mile-per-hour-plus supercar, Ferrari included 15-inch carbon-ceramic disc brakes. Around them are alloy wheels with Bridgestone Potenza RE050 Scuderia tires, the fronts being 245/35ZR-19s, the rears 345/35ZR-19s.

Carbon fiber is also a main ingredient in the Enzo's interior, looking as studly as it is structural. The red-faced tach and speedometer are hooded straight ahead, just past a very purposeful steering wheel. No sissy buttons on this wheel for radio volume or telephone activation, but big-time stuff, like shock absorber settings and traction control.

Ah yes, that supercar staple, the flip-up doors. When you open the Enzo's, it's surprising how light they are, until you consider that much of the car is fashioned in carbon fiber. Thanks to the lightweight material, Ferrari was able to hold the Enzo's curb weight to under 2,800 pounds. When introduced, the shape was controversial, but has since come to be admired by most supercar fans. It doesn't hurt, of course, that during the exotic Ferrari's production run, Michael Schumacher was nailing down win after win for the company's Formula 1 team.

KEN OKUYAMA ON THE FERRARI ENZO

Some of us love it and others don't, but it's tough to find anyone with a neutral feeling about the exterior design of Ferrari's Enzo. And then its designer reveals that the basics of the supercar from Maranello were drawn in fifteen minutes over a lunch hour.

Enzo lovers might say, "Eureka! A stroke of brilliance . . . fifteen minutes of genius."

Here's how it happened:

Ken Okuyama, a well-known Japanese designer who worked at Pininfarina, left to teach at Pasadena's famed Art Center College of Design and then returned to Pininfarina and penned the Enzo.

Okuyama smiles and begins, "Ferrari is such an emotional brand, so the decision-making process is always emotional."

Pininfarina had a proposed Enzo design ready to go, "which was much more like a Group C Le Mans racer," Okuyama explains. "The design made sense but didn't give us the heartbeat we wanted." As they approached an afternoon meeting during the car's design approval process, Okuyama explains, "About an hour before the presentation, we said of the Group C design, 'This is not going to go. We need something else.'

"I made a sketch during lunchtime of the concept of a Formula 1 car underneath four fenders, an F1 two-seater in plain view with four fenders. That was the concept of the car. So the Enzo was sketched in fifteen minutes during a lunch break.

"When we presented it after lunch, we had a mixed reaction from people, but my boss, Sergio Pininfarina, said, 'We're going to turn this into a model.'

"We did, and when we had a final presentation, the Ferrari engineers—about thirty people—went against the F1 direction because it was complicated, not Ferrari."

Luca de Montezemolo, president of Ferrari, was at the meeting and asked, "Which do you like?" Everybody went for the Group C car.

But then Montezemolo said, "Okay, but there's no democracy," and endorsed the Enzo.

In the end, only three men wanted the Enzo, but they were the most important: Montezemolo, Pininfarina, and Okuyama.

It proved to be a controversial decision, and the initial press photos released of the Enzo did it no favors. "I almost broke into tears when I saw them," Okuyama admits. "That's not the car. I had to tell people to wait until the car came out."

How did Okuyama react to the controversy over the Enzo? "We're used to it. That's just the case for car designers."

Okuyama does comment, "The only thing I don't like about the Enzo is that the roof was meant to be black. It has to be black to see the silhouette of a Formula 1 car. And it was black until the last minute, when it was changed to body color."

He also reveals that "we had two versions of the nose until the last minute, a sports car nose and a Formula 1 nose. We weren't sure if Ferrari was going to win the Formula 1 Constructors Championship that year, and if we'd lost the championship, a Formula 1 nose wouldn't have been right. But we won that year [1999] and chose the Formula 1 nose."

He adds, "I don't think we should take the same approach for all production Ferraris. They should be more elegant and easier for most people to swallow, but in the Enzo's case, we really wanted to make a statement for decades."

They certainly did.

The business end of the Enzo: a 65-degree aluminum V-12 that dynos up to 650 horsepower at 7,800 rpm and 485 lb-ft of torque at 5,500. Thanks to its Bosch engine management system, the big engine will not only go into afterburner mode but will also happily putter along in traffic when needed.

The leather-covered carbon-fiber shell seats hug the driver around the middle for lateral control, without squeezing inappropriately tight.

The most important thing in the interior is the big red start button that fires the 6.0-liter twin-cam forty-eight-valve V-12. With a mighty 108.4-horsepower per liter, the 65-degree V-shaped aluminum powerplant comes in at 650 horsepower at 7,800 rpm and 485 lb-ft of torque at 5,500. What makes the Enzo so eminently drivable in city traffic or on the open road is the Bosch Motronic ME7 engine control system and the fact that you have some 380 lb-ft of torque down around 3,000 rpm.

So you could putter your way to the store with little concern. When you want power and push the accelerator, the intake stacks will literally shorten to change the intake runner length, and the V-12 will instantly respond.

It was difficult to resist doing this every quarter-mile or so when driving the Enzo. Hard to get into the touring mode as we drove from Provo to Newport Beach, because the Ferrari is so easy to drive and, as a result, so tempting. We could be at 100 miles per hour lickity-split. A bright red Ferrari Enzo was also the main object of attention to anyone else (this would include the police) on the road. We were the Martian in a grocery store.

Naturally, the Enzo drew a crowd whenever and wherever we stopped. Gassing up the Ferrari proved to be an adventure, the car causing so much commotion in one station that a young lady backed her Toyota into a phone pole in the confusion.

When sold new in the United States, the Enzo was government-certified with the worst fuel mileage in the country at 8 miles per gallon in city traffic and 12 miles per gallon on the highway. In fairness, that's about the same as a Dodge Ram pickup truck, but the Ferrari is much more fun. Those who bought Enzos new paid $7,700 as a "gas guzzler tax," a pittance compared to the car's $652,830 sticker price, which, by the way, included a three-piece set of fitted leather luggage.

Speeding ticket-free back in California, we were anxious to get the Enzo to the test track and see if its true numbers matched our on-the-road enthusiasm. In the 700-foot slalom, the car slithered through at 73 miles per hour, fastest ever for *R&T* with the steering and, in test driver Patrick Hong's words, "quick and smooth, as though responding telepathically." Around the skid pad, the Ferrari generated 1.01 g.

After a few runs getting the launch control down pat, acceleration tests began.

Off with ASR, put the car in race mode, left foot down on the brake, right foot sets the throttle—about 2,100 rpm seems best—then step off the binders. Like Michael Schumacher blasting off a Grand Prix grid, the Enzo does an electronic launch, the driver's main responsibility being to limit wheelspin with the throttle. The needle arcs quickly toward the redline. Don't lift, pull on the upshift paddle, and, in 150 milliseconds, you're in second . . . then third . . . then fourth . . .

Get this right and the Enzo is at 60 miles per hour in 3.3 seconds, at 100 in 6.6 seconds, and through a quarter-mile in 11.1 seconds at 133 miles per hour . . . 0.5 seconds and 8 miles per hour faster than a McLaren F1. And with an estimated top speed of 218, it's nice to know the Ferrari also set a new *R&T* braking record by stopping from 80 miles per hour in 188 feet.

At this point in the testing, even the most understanding of owners would be suggesting we've got the numbers, how about dinner?

Not Richard Losee. He encouraged Patrick to try more acceleration runs. Who was the tester to offend the owner? The Enzo was kept at it all day and never so much as spit, coughed, or leaked one drop of oil. The next day Losee calmly filled the gas tank and drove back to Provo.

So much for temperamental exotic cars.

Stung by criticism that it built too many F40s and diluted the pleasure

(and profitability) of owning the exotic car, Ferrari promised it would build 399 Enzos and no more.

They lied by one, but for a good cause. After the production run in 2002–2004, Ferrari assembled one last Enzo and gave it to Pope John Paul II. It didn't fit in the Vatican motor pool, and this last car was auctioned for $1.2 million, with the proceeds given to charity.

That price is Bugatti Veyron 16.4 money, but somehow, despite its value, an Enzo isn't as intimidating to drive as the German-engineered car and has a lighter, freer spirit.

It's almost more enjoyable in a good-time-was-had-by-all sense. Like the time on our Provo-Newport Beach trip when I was driving the chase car behind the Enzo and we were doing 70 miles per hour on a dry, straight, level, beautifully paved stretch of pavement way out in the desert. No one in sight. The Enzo's driver planted the gas and, from my perspective, the car quickly shrank from full-size to red nano-particle. I won't tell you who was driving or where it happened, but 185 miles per hour was seen.

Wow. Your mother wouldn't do that . . . would she?

A picture is worth 1,000 words, and an Enzo is worth $652,830—at least it was when new. Expect to pay more than $1 million to buy one today. A $7,700 lump in the original US purchase price went toward the government's "gas guzzler tax."

Ferrari evolved the Enzo into the FXX, which began deliveries in early 2006. The V-12 engine is enlarged to 6.2 liters, power punted from 650 to 800 horsepower. It's said that at $1.8 million, the FXX is the most expensive new car ever marketed. You had to be an established, accepted Ferrari owner to buy one of the twenty-nine made.

Looking low and delightfully nasty, Maserati's midengine MC12 marked the return of this historically important marque to the supercar set. The white-with-blue-trim paint scheme on the V-12-powered exotic is taken from the company's roots, being the colors used by "Lucky" Casner's America Camoradi Scuderia race team, which campaigned Maseratis.

Maserati MC12
2004

205^{mph}

Maserati's long, white supercar snarls just a bit as it runs up on me, looking as though it might bite. The MC12 would love to be unleashed, to leap ahead and disappear, to let that 630 horsepower go and speed on to its 205-mile-per-hour top velocity. Considering it can slam to 60 miles per hour in just 3.7 seconds, that probably wouldn't take long, the driver blipping through its six-speed sequential gearbox.

Right now, however, it's picture time.

Automotive photographers spend a lot of time in the open trunks of cars looking back at exotic machines that are chasing them for a photo. It's called a car-to-car and the scene is spectacular.

Imagine Mario Andretti in his Formula 1–winning Lotus, its nose just inches away at speed. Phil Hill in a snarling Ferrari Testarossa 2 feet back. Brian Redman chasing your tailpipe in the 1970 LeMans-winning Porsche 917.

Or Maserati's MC12 as we circulate around Ferrari's Fiorano test track. Okay, we aren't at its top speed, but that doesn't take away from the image as the supercar chomps at our heels.

It's encouraging to see Maser's famous trident symbol on the nose of a lightweight midengine machine meant to drub the likes of Saleen and Aston Martin on the world's race circuit.

It has the pedigree on two fronts. In racing, Maserati was a major Grand Prix threat until the late 1950s and a thorn in Ferrari's racing side until the early 1960s. The MC12's paint scheme—white with blue trim—echoes the US racing colors used by "Lucky" Casner's America Camoradi Scuderia team.

Under its carbon-fiber bodywork is another pedigree: its basic tub inherited from corporate cousin Ferrari, the one used in the Enzo. Ditto with the 6.0-liter, 65-degree V-12 and the paddle-shifted six-speed gearbox.

This is not, however, an Enzo clone.

For one thing, it's bigger. The MC12's wheelbase is longer and its overall length, width, and height of 202.5, 82.5, and 47.2 inches make it, respectively, 17.4. 2.4, and 2.2 inches larger than the taut Enzo.

American Frank Stephenson guided the Giorgetto Giugiaro design of the MC12 through the wind tunnel and final development process, striving for beauty and downforce. It's a striking shape: the low, wide nose; the beautiful grilles over the outlets on the hood; louvers along the fender tops and the engine cover; the sweeping sculptural raised rear wing; plenty of front and rear overhang; and a long, unmistakable presence. Plus a few more pounds than the Enzo, which weighs in at 2,766 pounds versus the MC12's 2,943.

Beautiful and, unlike the Enzo, done as a removable-top Spyder.

Where the Enzo has the strong taste of carbon fiber and racing inside, the MC12 is more completely finished like a road car, though just as functional as the Ferrari. Here, carbon fiber is used more as visual trim, playing second fiddle to leather, a dash cover called BrighTex, and aluminum. Another major difference: The start button is blue.

Behind the cockpit is the V-12, which comes in at 630 horsepower at 7,500 rpm and 481 lb-ft at 5,500, so it's slightly detuned from the Enzo's 650 at 7,800 and 485 lb-ft at 5,500. Other than that, the V-12 is as seen on the Ferrari, with twin cams working four valves per cylinder, plus dry-sump lubrication.

It's an Enzo ditto again with the upper and lower A-arm suspension and pushrod-actuated inboard springs and dampers. The Brembo disc brakes measure 15 inches in front and 13.2 inches in rear, with ABS and electronic brake force distribution.

Kim Wolfkill, editor of *Road & Track*'s, had a chance to test the MC12 and wrote,

> An effortless handler at normal velocities, the MC12 comes alive as the pace climbs and cornering loads increase. Like any good race car, the big Maserati offers more feedback the harder the chassis is pushed, responding with increased levels of grip control and steering feel. Any previous hint of corner-entry understeer disappears with increased chassis loads. Mid-turn behavior is always rock-solid and wagging the tail in anything

Porsche's automobiles have won thousands of races since 1949, ranging from minor club events around the world to the most important competitions on the planet. Here a Porsche 911 RSR leads a Ferrari 488 GTE in the infield of the 2018 Rolex 24 at Daytona. Unlike earlier Porsche race cars, this one is midengined, its 4.0-liter flat-6 mounted ahead of the rear axle.

Maserati's MC12 is based on the same basic carbon-fiber monocoque tub as the Enzo of corporation cousin Ferrari. While it is not a clone of the Enzo, the MC12 also has the same upper and lower A-arm suspensions with inboard springs and shocks operated by pushrods.

Where the Enzo's interior is carbon-fiber race car efficient, the MC12's is finished more like a road car. Leather and aluminum are the main themes, with carbon fiber a lesser element. While many supercars have a red START button, the Maserati's is blue. Unlike the coupe-only Enzo, the Maserati has a removable hardtop for semi-open-air driving.

but first- and second-gear turns requires a deliberate effort to back the car off the road. Once into triple digits, the downforce contributes further to the car's grip, adding an unusual element of faith to negotiating fast sweepers.

Maserati had to build twenty-five MC12s to meet racing regulations, but did fifty over the course of two years, all presold. Not to the United States, though a few were imported.

Although racing was the MC12's purpose, something got bollixed up and the car ended up too long and wide to meet the rules under which the 24 Hours of Le Mans are run. Maserati shortened the car's length so it could run in the American Le Mans series, but it couldn't make it narrower, so it had to run for no points. However, in the FIA's cutthroat European GT series, the Vitaphone team won the 2005 GT1 Championship with MC12s.

It certainly looks like a winner seen down low in my photo session, though also like a caged animal at this speed. With my Canon's memory card full, I waved the test driver by as we entered the straightaway at Fiorano. He grinned, downshifted one gear, and simply disappeared.

Zowie-e-e-e.

Road & Track hustled a Carrera GT to 60 miles per hour in 3.6 seconds. Tester Patrick Hong wrote, "Exiting the corner is when the driver has to be patient to unleash all of the V-10's might. Too eager on the throttle and the rear will step out."

Porsche Carrera GT 2004

205^{mph}

Like a jet fighter just before takeoff, a Porsche on the Mulsanne Straight at Le Mans is an awesome sight. Especially at night when you are only 20 feet away, you can feel the car fly by. Could be a 917, a whistling Turbo, a 956, a 962, or a GT1, but the speed could be near or above 200 miles per hour . . . about as fast as your heart is beating after seeing it.

Porsche and the 24 Hours of Le Mans have been a dynamic duo for decades, so it's appropriate that the German automaker's supercar, the Carrera GT, has its roots there. Almost raced there.

But our story begins in another French city: Paris.

Just about the last thing you want during the Paris Auto Show is a 6 a.m. unveiling (ugh!), but Porsche promised it would be worth the early wakeup call. This was in autumn 2000, and it was dark and rainy when we trudged to the glowing pyramid of the Louvre, much too early . . . but it proved well worth it.

There, in the home of great art, Porsche presented its own masterpiece: the Carrera GT.

Masterpiece—is that a bit melodramatic?

Not really. Many automotive journalists and enthusiasts were still ticked off that Porsche announced it would be building the Cayenne sport utility vehicle (SUV), and we needed an indication that the famous automaker from Zuffenhausen hadn't left its moral compass untended. The Carrera GT was proof Porsche hadn't lost its way.

Porsche figures that automobiles with its name have won some twenty-three thousand races since 1949, so it came as no surprise to learn that the Carrera GT was originally meant to be a competition machine.

It was 1998 and the German automaker had just won the 24 Hours of Le Mans with the ground-hugging GT1. Since Dickie Attwood and Hans Hermann won the French classic with a 917 in 1970, race cars with the name Porsche have won sixteen times. In the afterglow of that victory, the thought was to create a roadster successor that would continue the Porsche run. Corporate priorities would derail that plan, but instead of the concept ending up abandoned on a siding, it was rerouted to the production department.

Run down the specifications of the Carrera GT and it is easy to see that it was conceived as a race car.

Begin with low weight. It isn't unusual to find an exotic car with lightweight carbon fiber used anywhere from bodywork to monocoque components. Porsche, on the other hand, uses carbon-fiber-reinforced-plastic (CFP) for the entire monocoque structure tub of the Carrera GT. Think of that tub as every bit of structure that surrounds you when you sit in the car: behind you, to your sides along the doorsill, the entire area in front of you, and even the windshield surround above you.

This is the basis for the Carrera GT. An impressive piece and extremely strong, the tub weighs only 220 pounds and is the first use of a single CFP chassis structure in a road car. Needless to say, it is also a very expensive piece, so if you are going to crash your GT, *don't break the tub!* The repair bill will match at least half the cost of your $440,000 Porsche.

Porsche doesn't stop there. Bolted to the vertical rear wall of the tub is another CFP structure Porsche calls the engine carrier. So the GT is basically a very strong, impressively stiff carbon-fiber chassis from nose to tail. The CFP doesn't rust, isn't affected by other environmental stresses, and looks impressive, which is why Porsche lets the carbon-fiber surface show through in several places on the car, like on the dashboard. Also made of carbon fiber are the doors, front and rear decklids, the doorsills, and the rear side panels, and if you look very closely at them, you can see just a hint of the cloth's pattern. Ditto with the removable roof panels that weigh only 5.3 pounds. Adds to the image.

Almost taking away from the image, when you close the lightweight doors, you don't get the solid "thunk" you'd expect, but what's best described as a lighter-weight sound.

Light weight was also the point of the design of the Carrera GT's engine, but so too were power and an impressively low profile.

Tipping the scales at only 452 pounds, the V-10 is made of aluminum, with its coolant and oil passages integrated into the block. Keeping weight down means more than creating an engine that isn't heavy; it also means an engine with lower reciprocating mass inside the V-10 so it will rev freely—hence the three-ring aluminum pistons, the titanium connecting rods attached to the forged crankshaft, and a redline of 8,400 rpm.

The longitudinally mounted V-10 has an unusual 68-degree vee angle separating its cylinder banks. There are four valves per cylinder, with the intake camshaft chain-driven (plus VarioCam), which is Porsche's own system of variable valve opening.

An aluminum intake manifold feeds each of the engine's banks, and just as the free-revving 5.7-liter V-10 hits 8,400 rpm, it is producing 605 horsepower or 105.5 per liter. Torque tops out at 435 lb-ft at 5,750 rpm, though it already hits a nice plateau of power around 4,500 rpm as it continues to climb.

For a competition or high-performance car, it's important to get the drivetrain as low as possible to keep the center of gravity down. The first step in this process is traditional—designing the V-10 with a dry-sump oil system. So instead of the usual reservoir of oil sloshing around in the crankcase below the engine, it is pumped to the engine for lubrication, then collected at the bottom of the V-10. Nine pumps then move the lubricant to an external reservoir, which in the Carrera GT's case is in the gearbox housing. Using the dry-sump design also means the engine

Porsche figures its automobiles have won some twenty-three thousand races since 1949, ranging from minor club events around the world to the most important competitions on the planet. After winning the 1998 24 Hours of Le Mans with a GT1 like the car seen here, Porsche considered building a new roadster race car, the design that became the Carrera GT.

Weighing just 452 pounds, the Carrera GT's 5.7-liter V-10 has an odd spacing of 68 degrees between its cylinder banks. A very free-revving engine, the V-10 is rated at 605 horsepower at 8,000 rpm and 435 lb-ft of torque at 5,750 rpm. *Photo by Marc Urbano, Road & Track*

Porsche's Carrera GT supercar was originally intended to be a race car, but when corporate priorities changed, the design was turned into one of the hottest supercars on the street. A testimony to lightweight design, the Porsche has a top speed of 205 miles per hour.

is more efficiently lubricated under extreme driving conditions, such as high side load corners.

That is only one design feature of the Carrera drivetrain that keeps the height and weight low. Even with a dry-sump engine, a driveline's height is often kept higher than desirable by the flywheel and clutch diameter. Add still another Porsche patent for the Carrera GT, this one being the ceramic composite clutch. Engineers managed to slim the flyweel and clutch dramatically. As an example, the 911 Turbo's clutch weighs 15.4 pounds and measures 15 inches in diameter, yet the Carrera GT's clutch weighs just 7.7 pounds and measures 6.6 inches. With clutch plates made of a ceramic composite said to be extremely durable, this design not only drops the engine to just 3.9 inches above the floor of the car, but the flywheel-clutch package's lower rotational mass allows the V-10 driveline to rev that much more freely.

Porsche opts for a traditional six-speed manual; it is quite compact and is located transversely behind the engine. Its casing also houses the limited-slip differential and the V-10's starter motor.

Going back to its Le Mans roots for the Carrera GT's suspension, specifically to the 911 GT1, the GT has upper and lower aluminum A-arm suspension front and back; the shock absorber and spring units are actuated away from the A-arms via rockers and pushrods. This design not only aids in setting up the suspension but also gets the weight of the shocks and springs away from the wheels. Steering is through an assisted rack and pinion mounted to the front bulkhead.

Porsche's ceramic composite disc brakes, which are only half the weight of conventional metal brakes, also aid in minimizing unsprung weight at the wheels. These are less susceptible to environmental effects and are said to be quite durable. With a diameter of 14.96 inches, the discs are cross-drilled and vented to get the heat and moisture out and are stopped by six-piston calipers.

Grant Larson, the American designer who penned the Boxster, also gets credit for the Carrera GT. It is pure Porsche, from its 917-inspired headlights to the spoiler-topped tail. Its compact proportions are 3.5 inches shorter than a Ferrari Enzo and almost an inch lower.

Porsche builds two vehicles in its new Leipzig factory: the Cayenne SUV and the Carrera GT. This is not a manufacturing facility for the sports car, but an assembly line. Components such as the tub, engine, and suspension are built off-site and brought together to create finished cars in Leipzig.

ABS and traction control are part of the design, and the latter can be disabled if you dare. Michelin supplies the GT's tires, Pilot Sport 2 265/35ZR-19s at the front, 335/30ZR-20s at the back, mounted on wheels made not of the expected aluminum alloy but of magnesium, again to keep weight down.

You can't see much of this, because it is hidden under the wonderfully aggressive bodywork designed by American Grant Larson, who also drew the original Boxster. It's hard to say which angle is the most exciting from which to view the car. Some would argue the front end, with its typically large Porsche inlets and Bi-xenon headlamps, their shape said to be inspired by the 917.

"I love it. I like the fact it's a modern car that pays homage to people who like to shift." And then his love of technical qualities comes out when Jay Leno adds, "I bought it primarily because of the clutch.

"The Countach and McLaren F1 are wonderful cars, but if you drive them the way you are supposed to, you burn the clutch up. In the McLaren, the clutch is gone at 2,500–3,000 miles. With the Countach, you pull up to the light and there's the kid with the hot-rodded VW Jetta or whatever. He's going to blow your doors off if you don't want to see the Countach clutch too hard. You can do maybe three or four burnouts and you're screwed.

"With the Porsche, you've got that little 6-inch ceramic clutch; it takes a tremendous amount of force and you can do burnouts all day long.

"When we were at Talladega [the high-speed NASCAR track where he was involved in high-speed record runs], we dropped the Carrera GT clutch at least fifty times just trying to launch it. It didn't complain, it didn't bitch and moan, didn't smell. It was amazing, so I like the fact a lot of effort was put into something you don't necessarily see.

"When I watch these shooter tune-up shows, it's always Porsches and Corvettes because they're the strongest. People work on them themselves, they beat the hell out of 'em. I never see a Lamborghini or Ferrari at one of these tuner things because the cars can't take the abuse the Germans and the Corvettes can. So that's what I like about the Carrera—you can really drive it hard."

The Carrera GT does have its downside, so to speak: "It's really low. You can't pull into any sort of driveway. I have the factory lift kit in it but can't take the Porsche home because it won't fit in the driveway. If you're getting gas, you're going, 'I know I'm out of gas, but I can't go to that station because there's a lip on the driveway'"

Not that it's the sort of problem you should have every day: "I met a guy who was complaining about his Carrera GT. He used it every day in San Francisco traffic. Well, you don't drink champagne every day unless you go crazy. These are weekend things . . ."

Then there's the side view with the huge front vents that duct air from the front radiators and brakes, leading it back to the nicely sculpted intakes for engine air. This side view is where you see the GT's different proportions, the best example being the fact that it is 3.5 inches shorter than a Ferrari Enzo but rides on a wheelbase that is 3.2 inches longer. Only 0.7 inch lower than the Enzo, the GT's body is 4.5 inches narrower, though their tracks are only about 3 inches different.

Arguably the best view would require you to climb a 6-foot ladder located just off a rear corner of the car and look down on the GT, with its mesh intake manifold covers that run forward to just short of the steel-reinforced rollover protection hoops. There's the rear spoiler with the center section that rises at speed, huge taillights, and check out the huge exhaust outlets on either side of the license plate.

Naturally, the shape's aerodynamics were verified in a wind tunnel, where the airflow under the body was also refined to add downforce.

When you get into a Carrera GT, there's no mistaking it for a Porsche. The starter key is on the left, a remainder of racing tradition. There are five dials straight ahead, stylized but classic Porsche nonetheless. You are sitting in seats on lightweight carbon fiber and Kevlar shells, finished in leather and offered in two different butt widths.

A rising center console divides the cockpit in this driver-oriented layout, and while the wood inlay knob shifter looks as though it's situated a bit high, its location turns out to be about perfect for driving, being just to the right of the steering wheel. All this is beautifully fitted with leather, metal, carbon fiber, or paint surfaces.

You can order your Carrera fairly stripped or add from a list of no-cost options, such as air conditioning or a navigation and Bose sound system. In any event, you get such genteel equipment as power windows, remote central locking, and a set of luggage made to fit the car's storage area and done in the same leather as the interior. Would you expect anything less?

Despite those amenities, Porsche's weight-saving work keeps the Carrera GT's weight to just 3,042 pounds. The light weight, combined with V-10 power, enabled *Road & Track* to get a Carrera GT to 60 miles per hour in 3.6 seconds, which is what you expect from an exotic in this class. Top speed is 205 miles per hour. On the skid pad, the Porsche managed just 0.1 shy of the magic 1-g level, which is impressive. Test driver Patrick Hong said, "Exiting the corner is when the driver has to be patient to unleash all of the V-10's might. Too eager on the throttle and the rear will step out. Even then, the onset of oversteer is so progressive that it will never catch you by surprise."

Porsche will tell you that without the profitable Cayenne SUV, developing models such as the Carrera GT wouldn't fit in the profit plan. Appreciate the Cayenne or not, it's somewhat ironic that not only are the pair of Porsches the only products of a new factory in Leipzig (part of the former East Germany), but also the 521-horsepower Cayenne Turbo S is Porsche's second-most-powerful vehicle after the Carrera GT.

The Leipzig factory has a sole, snail-slow Carrera GT line next to the Cayenne's, which you can actually see move.

Not so through the eight stations in which the GTs are slowly assembled. And "assembled" is the key word, as nothing for the car is actually manufactured here; it's put together from parts built elsewhere and shipped in. The assembly line is a fantastic place to see the building blocks of the GT, beginning with the carbon-fiber main section that has the tubing of the forward crash system jutting out front.

Along the stations, you see the engine with its carbon fiber added, as well as the suspension systems, the carbon ceramic brakes, the interior, and, finally, the exterior body panels. After being finished, each Carrera GT is test driven on a 1.1-mile circuit that is part of the 2.3-mile, FIA-approved, Formula 1–quality racetrack Porsche built next to the Leipzig factory.

Watching a Carrera GT whip through a replica of Laguna Seca's Corkscrew turn and approach the slow right-hander that leads onto the back straightaway, it's impossible to not wonder what the midengine Porsche would look like, numbers on its sides, wide-open exhausts howling away in pure competition form.

After all, that's what it was meant to be in the first place.

This cutaway shows the technical details of the Carrera GT. The 605-horsepower V-10 is mounted behind the cockpit. At its center is a monocoque structure made of carbon-fiber-reinforced plastic, a cell that surrounds the driver and passenger. Weighing only 220 pounds, it has the engine carrier mounted behind it and front suspension and crash structure ahead. Porsche cut unsprung weight at the wheels with ceramic composite disc brakes of its own design, which weigh half as much as conventional discs. Drilled and vented, the discs are stopped by six-piston calipers. *Drawing courtesy Porsche North America*

Ford was looking for a symbol it could use to celebrate its hundredth anniversary in 2003 and figured a modern interpretation of the Ford GT would do the trick. Just a concept car at the Detroit Auto Show in January 2002, the GT was engineered quickly; a trio of cars was ready at the anniversary ceremony in June 2003. Everyone knows the story—Henry Ford II was determined to beat Ferrari at the racing game in the '60s. This obsession resulted in the Ford GT40. After a disastrous first few years, Ford succeeded in beating the Italian automaker, winning the 24 Hours of Le Mans from 1966 through 1969.

Ford GT
2005

190^{mph}

To understand the excitement generated by Ford's GT, it's crucial to understand how important Ford's original GT40 was to US pride. American automobiles couldn't make it big on the international racing scene after World War II. Briggs Cunningham came very close to winning the 24 Hours of Le Mans in the early 1950s, but Americans just couldn't seem to get a major victory outside the United States.

American drivers had succeeded big time in Europe—Phil Hill, Dan Gurney, Richie Ginther, Carroll Shelby, Masten Gregory, and others—but not American cars.

After the first few years of Ford's GT program at Le Mans, things didn't look much better. Losing again in 1965, Henry Ford II basically told the men in charge of that racing program (including Carroll Shelby) that if they didn't win in 1966 they would be looking for new jobs.

There was no budget limit, and Ford went on to finish 1-2-3 with the GT40 Mk II in 1966, winning once more with the Mark IV in 1967, a truly all-American victory with Dan Gurney and A. J. Foyt driving. When the FIA cut back engine sizes for 1968, effectively turning the Mk IVs into museum pieces, wily John Wyer resurrected the small-block GT40s, and one of his machines—chassis 1075—won in both 1968 and 1969.

Some have argued that the GT40s were an Anglo-American effort; they were correct up to a point, but the program was covered with Detroit money, and the Mk IV was all-American.

So when Ford unveiled a GT40 concept car at the January 2002 Detroit Show, Americans felt immense pride in the famous old design's resurrection.

The following month, Ford held a "skunk works" meeting at Steve Saleen's factory in Irvine, California, where Saleen converts standard Mustangs into high-performance road cars. Engineers had to determine the feasibility of producing the GT. Could it be done economically? Did it make any sense?

"That's an emotional thing for me because when I was a kid we were a Ford family. When I got my dad to buy the 7-liter . . . we had Falcons and my dad had a '64 Galaxy, so we went in to buy the '66 and I said, 'Dad, can I pick the engine?' My mother was saying, 'Oh, let the boy pick the engine. What difference does it make what engine is in the car?'"

"So I met with the salesman, Tom Lawrence, and I pulled him aside and said, 'We want to get the Galaxy with the 7-liter option, the C6 automatic, and the glass-pack mufflers.' So we wait about four weeks and the car comes in and my father and I go to pick it up. He starts it and it goes: 'Hunga, hunga.' The salesman says, 'Mr. Leno, you didn't order mufflers. You wanted the glass packs.'"

"And then, when we leave, my dad pulls out of the dealership, steps on the gas, the car fishtails with all that power. My dad says, 'Jesus Christ, what did I buy?'"

"I remember a year later I found, in a drawer in a bedroom closet, a ticket he got for going 110 on a trip to Indiana."

"I was one of those kids who didn't care about the Super Bowl, but I remember seeing Ford win Le Mans. Americans have this wonderful sense of being underdogs. Especially in the '60s: we were the most powerful country in the world, we won World War II, we produced 15 million automobiles a year, and yet we beat a little Italian company and somehow we see ourselves as the underdog fighting. It makes no logical sense. But I remember the GT40s coming in 1-2-3 (in 1966), and I really believed there was a connection between that car and the Ford Falcon my mom drove and the 7-liter my dad had."

"Plus, I always think it's the most perfect sports car. You know, when a Jaguar XKE goes by, people who know nothing about cars—even in New York City where they don't care about cars—people say, 'Hey, look at that. That's something, isn't it?' and the GT40 is like giving a bath to a beautiful woman. There are all these curves and places. You don't get that with a Countach. You get it with the Ford GT40, the XK120, especially the Miura, any of these cars."

"The Ford GT was a car I just had to have when I heard they were going to build it. You know, it's only 13 percent bigger than the original and, hey, if I was only 13 percent bigger than I was in '66, I'd be thrilled."

After the California meeting determined that the midengine sports car could be built, Ford executives gave it the green light, but with the condition that three GTs must be ready for the celebration of Ford's one hundredth anniversary in June 2003. It was, to say the least, a crash program made possible because so much of the work—from exterior shape to chassis design to crash simulations—could be done with computers.

Although Steve Saleen created his own midengine exotic car (the California-built S7), and his company assembles the Ford GTs in a factory near Detroit, the two supercars share few roots.

One exception: Neil Hannemann, chief engineer of the S7, was deputized to Ford to be chief program engineer of the GT. He impressed everyone that his next job was executive director of engineering for McLaren Cars in England.

The new Ford GT also had little in common with the original GT40, though they both are powered by Ford V-8s mounted behind the cockpit and have their exterior design similarities.

In explaining the retro look of the GT, J Mays, who runs Ford's design efforts, pointed out that they tried to do a modern interpretation of the GT40, but it never looked quite right. Doug Gaffka, who led Ford's Living Legends Studio, explains, "The bottom line is, if you're doing a Ford GT, it had better look like a Ford GT."

The new midengine supercar's exterior, designed by good guy Camila Pardo, is 18 inches longer and 4 inches taller than the original, which was called GT40 because it measured 40 inches tall. Ford was not tempted to call the new version the GT44 and had to abandon the idea of using GT40 when the man who now owns that name wanted too much money for its use, so the automaker simply settled on Ford GT.

A short front overhang might be popular today, but keeping the original car's proportions meant the GT had to have a long nose, which proved

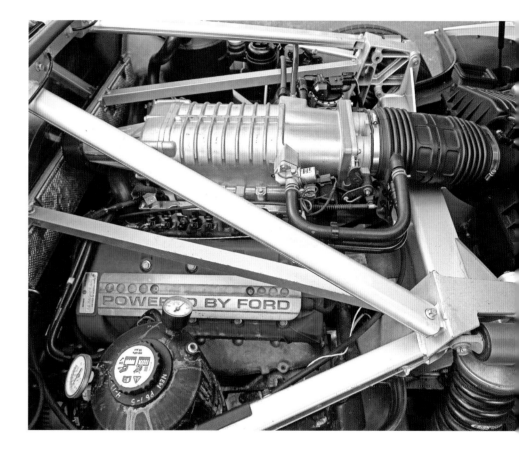

 Ford adapted the supercharged 4.6-liter aluminum V-8 from the SVT Mustang Cobra for the GT. Displacement was upped to 5.4 liters for the thirty-two-valve engine, which has a water-to-air intercooler and two fuel injectors per cylinder. Horsepower and torque both measure 500, about the same power as the 7.0-liter race V-8s from the 1960s.

handy for hiding the necessary bumper structure. Out back the flip-up tail spoiler was complemented by a small "floating" bumper that looks appropriate. Ford put the GT in a wind tunnel to tame the aero, particularly high-speed lift, and came away giving it a modern venturi tunnel underbody.

It may be a modern interpretation of a classic shape, but the GT is quite dramatic and exciting—not as soft, sweeping, and sophisticated as a Pininfarina Ferrari, perhaps, but appealing more to your muscles, to your gut.

There is another modern interpretation of the GT40 inside, from the basic layout of the instruments to the seats' grommet vent holes. What catches your eye first is a center tunnel of brushed magnesium that houses the fuel tank. On each side of it are leather-covered carbon-fiber shell seats. The console has stylized climate control knobs, and a row of modern-looking toggle switches is on the instrument panel,

J Mays, who runs Ford's design efforts, explains that attempts to create a modern interpretation of the GT40 just didn't work, and they reverted to the original shape. Designer Doug Gaffka further explains, "The bottom line is, if you're doing a Ford GT, it had better look like a Ford GT."

Ford's goal for the GT was to beat Ferrari's 360 Modena, which was a worthy target . . . and a historical one, as Ford and Ferrari were great race rivals. It can outrun the 360 to 60, and both have top speeds around 190 miles per hour.

Like many modern supercars, Ford's GT has a heart and soul of aluminum. In addition to the V-8 engine, the chassis is assembled from aluminum—extrusions, castings, and stamped panels. Around all this is aluminum bodywork with panels that are superplastic formed, heated to near 950 degrees and formed over dies with high air pressure.

as well as that most important of all controls, a big red start button for the supercharged V-8.

Plus, of course, there is equipment never dreamed of in a GT40, such as a CD-audio system, air conditioning, power windows, and a rear window defroster.

To get the GT's dramatic lines, particularly the doors that cut deeply into the roof like those on the GT40, Ford opted for aluminum exterior panels that are super-plastic formed, heated to almost 950 degrees Fahrenheit and then formed over a die with high air pressure.

Aluminum is also used for the space frame, a collection of extrusions, castings, and stamped panels that Ford claims is stiffer than the competition, the Ferrari 360 Modena and Lamborghini Gallardo.

Aluminum is found again in the unequal-length suspension control arms, which are matched with coil-over shocks and front and rear anti-roll bars. Steering is rack and pinion. The brakes are vented, cross-drilled, four-piston caliper Brembos with ABS, and the tires are Goodyear Eagle F1 Supercars measuring 235/45ZR-18 front and 315/40ZR-19 rear.

Ford already had a 4.6-liter supercharged 390-brake-horsepower twin-cam aluminum V-8 for the SVT Mustang Cobra, and that provided the base for the GT engine. In the midengine car, the V-8 is at 5.4 liters with four valves per cylinder, a screw-type supercharger, a water-to-air intercooler, two injectors for each cylinder, and 500 horsepower and 500 lb-ft of torque. As a gauge of technical progress, that fully emissions-certified engine has about the same power as the race-ready 7.0-liter V-8s in the 1960s GT40s.

Behind the low-mounted dry-sump engine is a newly designed Ricardo six-speed manual transmission with a twin-disc clutch and limited-slip differential.

Ford made the time goal, unveiling a trio of GTs—one each in red, white, and blue—at the company's centennial in June 2003. The number one production car was auctioned off that August to former Microsoft executive and Ferrari collector Jon Shirley, who paid $500,000 for the honor, though all the money above the car's retail price of just under $150,000 was donated to charity.

When getting into a Ford GT, it is important to remember that, as with the original GT40, you have to duck under the dramatic door cut-ins as the door is closed.

When I drove a GT40 race car a few years ago, I thought it would feel confined inside, but I was amazed at the wonderful, wide-open view through the windshield. Same goes for the new GT as you settle back into the seats, which have a vintage, reclined feel about them. You can tell this is a car that was well thought out, as everything has a natural feel to it and there is no hunting for switches or buttons. Dead ahead is the tachometer, right where it belongs. The supercharged V-8 rumbles right behind your head, more of a strong presence than a noise problem.

Since the GT's main rival was the Ferrari 360 Modena, just as Ferrari was Ford's big foe at Le Mans, it was important to beat that car to 60 miles per hour. And it does, the Ford getting there in 3.8 seconds, 0.5 quicker than the Ferrari, with both their top speeds being 190 miles per hour.

As important as the numbers, the Ford GT's power is wonderfully linear, building so strongly through each gear. The V-8 proves both powerful and flexible, just as happy to pull away from low revs in a high gear as it is to spin right up to its redline.

Even on a track, you can feel that this is a car with few vices for the average driver. There is the expected precise turn-in and enough sense of initial understeer to keep you comfortable in the car, in which the chassis seems matched to the power, neither overwhelming the other. And the brakes are willing to pull you down from speed repeatedly with no fade.

Getting out of the car, you again have to remember to mind your head, and you find yourself sneaking away while ducking too long to be clear of the door, the way people do when walking under spinning helicopter blades.

Ford has been holding production of the GT to around four thousand units, building some nine per day, which isn't enough for each of its dealers to get one. And some dealers who did get one wouldn't sell it, so soon after GT deliveries began, their prices climbed to as much as $250,000 when resold.

There are those of us who winced when we saw the GT40 concept at Detroit, dismayed that its design so closely copied the original. Couldn't Ford have come up with something new? Now, having driven the Ford GT and seen it on the streets, it's time to eat a little crow.

Ford also got historical inside the GT, bringing back the original GT40's grommet vent cooling holes in the seats and the basic shape of the dashboard. These days the seats have carbon-fiber shells, and there are modern climate control switches on the center console.

At the front of the GT, Ford was able to use the long overhang to build in crushability for crash tests. At the back, it created a floating bumper that does the trick and looks appropriate. In the wind tunnel, the fast Ford was detailed for aerodynamics, which included an underbody venturi tunnel. Aluminum is used for the unequal-length control arms in the suspension. The Italian firm Brembo adds the vented cross-drilled brakes.

If you're working at it, just 2.5 seconds after launch, you have the Bugatti Veyron at
60 miles per hour. Another 7.3 clicks of the clock and you've brushed past 125 miles
per hour, and within 10 more seconds (quick, take a breath), you're approaching
190 miles per hour. Stick with it under the right conditions and the sixteen-cylinder
engine will propel the 4,163-pound car to 250 miles per hour. Proven.

Bugatti Veyron 16.4 2005

250^{mph}

It was a recipe for disaster.

Heavy rain had fallen in the night, and there was little chance the roads that twisted though the tall hills of Alsace in western France would dry soon. It being autumn, the showers had laid down a heavy layer of gold and amber leaves, slippery as a greased anaconda.

Under us was 1,001 horsepower and, even with all-wheel drive, there's only so much we could do to hedge on the coefficient of friction. But Ditmar Hilbig had the pedal down and the Bugatti Veyron 16.4 slipped through the turns with remarkable speed and stability, given the slimy conditions.

He guided the Veyron into a hairpin paved with autumn leaves, and it remained stable and unperturbed. The left-hand wheels were on a drying surface, the rights on slippery leaves, but the Bugatti never quivered. He aimed the Veyron 16.4 through tight esses where the camber danced right-left-right, and the car never equivocated but stayed straight and true.

Magic.

It was difficult to tell if he was trying to impress or warn. He was driving on the outbound leg of this orientation drive, and I was to be in command on the way back. We were, after all, in a $1.2 million car . . . possibly worth more considering it was Veyron chassis number four. Ditmar's baby.

It's been a long time coming. Few cars have been as anticipated or through such a "when will they finish?" gestation period as the Veyron 16.4.

It is the fourth Bugatti created by Volkswagen since the German automaker acquired the famous name in 1998.

First came a trio of potential Bugattis, all shaped by Giorgetto Giugiaro's Italdesign. It began with the front-engine EB118 two-door coupe (1998 Paris Show), followed by the EB218 four-door sedan (1999 Geneva Auto Show), then the midengine W18/3 Chiron supercar (1999 Frankfurt Motor Show) built on the chassis of an Italian Lamborghini Diablo VT, another VW-owned automaker.

All three show cars had eighteen-cylinder engines with the W layout championed by VW and Dr. Ferdinand Piech, grandson of Dr. Ferdinand Porsche and godfather of such other renowned machines as the all-conquering Porsche 917 race cars—an engineering genius if not a subtle man.

It was under Piech's reign at Volkswagen that the company bought Bugatti, Lamborghini, and Bentley as its luxury brands. It then began the design and engineering of a Bugatti that would trump the supercar competition from the likes of Ferrari, VW's own Lamborghini, Porsche (on whose board Piech sits); plus the out-of-production-but-legendary benchmark McLaren F1.

Piech is a man who gets what he wants, and the Bugatti supercar is a prime example. The VW-designed car was first shown to the public with an eighteen-cylinder engine at the Tokyo Auto Show in late 1999 and in its definitive sixteen-cylinder form at the Paris Auto Show in autumn 2000.

The heart of the Veyron is its W-16, which is surprisingly light and compact. This last quality comes with the engine's layout, essentially a pair of dual overhead cam V-8s with only 15 degrees in the V angles set side by side and matched to a common crankcase. This sixty-four-valve 8.0-liter engine is assembled in Germany with such lightweight components as titanium connecting rods. With four intercooled turbochargers and direct fuel injection, the sixteen cylinders produce 1,001 horsepower at 6,000 rpm and 922 lb-ft of torque between 2,200 and 5,500 rpm.

It's a magnificent engine that looks the part . . . what you can see of it. It's possible to look down on top of the W-16 sitting behind the Veyron's cockpit, set like a huge diamond. The inside cam covers and intake manifolds are exposed to the elements, one of the concessions made to dissipate heat generated in making the 1,001 horsepower. But you must virtually disassemble the car to see any more of the W-16.

As technically interesting as this impressive powerplant is, its transmission is also awe-inspiring. The seven-speed gearbox has

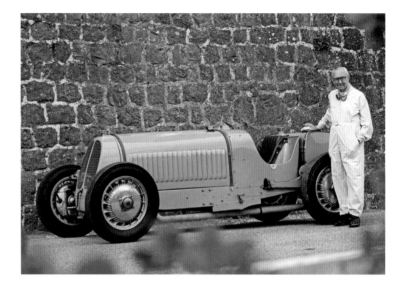

Ettore Bugatti created the Type 53 in 1931, making it the first major race car with drive to all wheels. The 5.0-liter engine and rudimentary AWD system made the car difficult to drive in circuit races, but it was a hillclimb winner. The late Rene Dreyfus, seen here with the Type 53, drove the car to a record at the La Turbie hillclimb in 1934.

Where a Ferrari might look like a lithe animal or athlete, the Bugatti Veyron reminds one more of a knight in armor. Behind the car is the stairway to Chateau St. Jean, part of the traditional Ettore Bugatti property in Molsheim in eastern France. Volkswagen has carefully restored the property and now uses it for meetings. A technical *tour de force*, the Bugatti Veyron features not just a 1,001-horsepower engine with seven-speed transmission, but also advanced ceramic disc brakes with six-piston calipers inside alloy wheels fitted with the widest production tires in the world, made by Michelin. The price of all that sophistication? A cool $1.2 million.

automatic shifting and a pair of clutches. Built in England by Riccardo, it is like a super-sized VW DSG transmission in that as one clutch disengages a gear, the second clutch immediately engages the next, making for extraordinarily smooth and quick gear changes. Although you can drive the Veyron in a conventional automatic mode, the manual shifting via paddles is usually too entertaining to pass up.

Even though the Veyron weighs a healthy 4,163 pounds, Bugatti logically figured that all-wheel drive was needed to get the power down with any sense of stability.

By Bugatti's figuring, 0–62 miles per hour comes up in 2.5 seconds; you're past 125 miles per hour in around 7.3 seconds. By the time you've counted "seventeen Mississippi," you're quickly approaching 190 miles per hour. The Bugatti's proven top speed, 250 miles per hour, makes this the fastest car in the world . . . which is what Piech wanted.

Suspension is the expected upper and lower A-arms with conventional springs and shocks, though the system has hydraulics that will raise and lower the car, depending on speed, and are part of the high-speed handling and aerodynamic package.

There are three modes.

Normal: Ride height is 4.9 inches front and rear, rear spoiler is down, front diffuser flaps open to let air under the car, and the drag coefficient is 0.393.

Handling (when manually deployed or above 135 miles per hour): Ride height drops to 3.15 inches front and 3.74 inches rear, the rear spoiler rises and can take an angle of 15–27 degrees for added downforce, diffuser flaps stay open, and the drag coefficient is 0.417. This is the maximum downforce setting.

JAY LENO ON THE BUGATTI VEYRON 16.4

"Bugatti says about the Veyron, 'Your wife could use this to go to the market.' Why? Why would I give my wife the Veyron to go to the market? Why would you even think of going to market? 'Oh, I've got the Veyron. I think I'll go to the market.' It doesn't make any sense to me," Jay Leno says.

"I got to drive it at Pebble Beach, and it's very impressive. Technically, it's amazing. It does suffer from that thing about not being able to see the motor, which for me is important. I mean, that seven-double-clutch transmission is brilliant but . . .

"Bugatti was more artist than engineer, and this Veyron is more engineer than artist. There's no sense of involvement with the car, nothing sort of rumbles, and you get in and press a button and you put it in 'D' and you go. It's a bit like being in a jet plane. The real fun part is the takeoff and landing, not the actual driving.

"I do appreciate the Veyron," Leno explains further, calling the engineering of the car, "a brilliant job, but physics are physics and it weighs 4,300 pounds. I always think of that book, *The Incredible Lightness of Being*, and I like that phrase. That's what I like about Lotus Elans.

"The dashboard is a little too 'bling, bling' for me. I come from the school of the airplane cockpit look. I like to see the oil temperature and everything. I find the gauges too small, but that might be more age than anything else. To me, knowing the oil or transmission temperature is more interesting than listening to the radio."

"Plus, I don't quite understand the car's purpose. It's not quite a GT, it's not quite a sports car . . . it does all those things well and it is fast. My great fear is that it's one of those cars that will bring prying eyes to the sport: '1,000 horsepower in this day and age! Can we afford . . .' Suddenly General Motors says we have to pull back on this and everybody decides. It just shines the light. That's the trouble. The articles I've seen in the mainstream press ask, '1,001 horsepower? Is this what we need now?'

"You know, the F1 McLaren obviously made a huge splash, but it made its huge splash among people who already knew about cars. It didn't really make waves among people who don't, and that's what I like. The idea that some idiot would go in and say, 'Hey, give me one of those Veyron things. Those are really fast, right?' And they hit a tree and the next thing there are calls for horsepower limits.

"That being said, it's a wonderful achievement. It's amazing."

High speed: Ride height lowers to 2.56 inches front and 2.75 inches rear, the rear spoiler retracts but maintains a 2-degree down angle, the diffuser flaps close, and drag drops to 0.355.

Incidentally, to get to that high-speed setting, you have to stop, turn a second key, and run through a checklist of very good ideas, such as making certain tire pressures are correct.

With the Veyron 16.4's acceleration and top speed, brakes have to be of a high order. Bugatti specified massive, high-tech carbon-ceramic brakes, the front discs measuring 15.7 inches and stopped by eight-piston, four-pad calipers, while the 15-inch rears have six-piston, two-pad calipers. Better yet, at high speed, a good stab at the brake pedal also tilts the rear wing forward up to 55 degrees to increase rear downforce and aero drag.

Tires threatened to be a problem. Bugatti needed them to hold together at 250 miles per hour, stay put at 1.3 g on a skid pad, and be run-flats, eliminating the need to stow a spare. A tough set of standards. Michelin came through with Pilot Sport PS2 PAX skins for the Veyron, claimed to be the widest production tires in the world, the rear treads taping 14.4 inches across, the wheels at 20 inches front, 21 inches rear.

That's the technical background, but what is it like to be around the Bugatti on its home ground in France, and then drive it?

Seeing the Bugatti outdoors, away from the auto show circuit, you understand why you don't find a lot of people head-over-heels in love with the Veyron's styling, though everyone comes away mightily impressed. While the Veyron isn't a beautiful car in the manner of a Bugatti Type 57 or Ferrari 250GTO, it is imposing and oh so interesting. Where the body of a high-performance Ferrari might remind one of an athlete, the Veyron is a knight in armor—Carl Lewis versus Ivanhoe.

Go ahead. Climb in and you will have little doubt you are about to drive the big stuff. You feel it around you, rather like a presence. Being a big

Inside the Veyron are comfortable, deep-set individual seats with a tall center console. Ahead is the hooded instrument binnacle with the speedometer reading to 420 kilometers per hour (261 miles per hour) and a horsepower meter reading up to 1,001. Note the nicely finished pedals and the low roof, the latter creating a slight "bunker" effect.

Giorgetto Giugiaro and his son, Fabrizio, with the Bugatti 18/3 Chiron show car, unveiled at the 1999 Frankfurt Auto Show. Built on the chassis of a Lamborghini VT, the Chiron was named for famous race driver Louis Chiron and was one of the prototypes for the production Veyron.

machine, it surrounds and envelops you, so there is a certain bunker feeling to being in the cockpit: low windscreen, high sides, the A-pillars a bit obstructive. The overall sense is imposing but not threatening.

The controls are elegant, as is the interior in the manner in which nicely finished metal elements—the Bugatti-esque engine-turned-console, the rimmed gauges, the door controls panel—contrast with high-quality leather. It's a look, an ambiance that isn't especially warm and enticing in the manner of a Jaguar or a Bentley, but it gets your attention and admiration.

Starting the engine isn't so much like starting a car as something more imposing (there's that word again), such as an unlimited hydroplane or a Spitfire (oops, Messerschmitt). Varooooom.

For all that drama, what happens next is quite civilized. Snick, away in first gear, the all-wheel drive planting the power on the ground, fast, secure, and stable. No snaking about, just the hand of God firmly in your back, pushing you forward—like a Gulfstream G4 on a high-performance takeoff, thundering toward the end of the runway. Then, snick, back on the upshift, and before you can take half a breath, you're up a gear, a new takeoff.

Where less-sophisticated transmissions might bang and slam you into the next gear, the Bugatti seems to instantaneously assume you into the following ratio, and off you go, so fast, so unobtrusive. Magic—again.

Before you say Ettore Bugatti, you're into third and the speedo is climbing to the right. Fourth . . .

Except this is someone else's $1.2 million machine, so I eased off the straightaway and the forest surrounded me. Quite pretty, but the wet roads were as treacherous as thin ice. I thought I sensed Ditmer relaxing.

I backed off and let the Bugatti assume its second personality. Because even with all its horsepower and torque, the Bugatti was not only super quick, but also dignified. Civilized.

Those $25,000 carbon brakes did squeal a bit under light braking, but they brought the Veyron to a dignified stop from speed quickly. The steering was light enough to make the Veyron quite maneuverable in town while still properly direct at speed.

In the end, the Bugatti is mightily impressive, but a bit overwhelming and not necessarily lovable in the manner of a Ferrari or that singular product of genius, the McLaren F1.

When you are with the Bugatti, however, there is the feeling that you are in the presence of greatness. Like being on a high-speed ride with Phil Hill when he's driving a vintage Ferrari. Attending an Eric Clapton concert. Or dropping into your airplane seat (as I once did) to find next to you is Walter Payton, who loved to talk about auto racing.

In a word, magic.

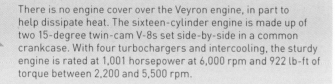

There is no engine cover over the Veyron engine, in part to help dissipate heat. The sixteen-cylinder engine is made up of two 15-degree twin-cam V-8s set side-by-side in a common crankcase. With four turbochargers and intercooling, the sturdy engine is rated at 1,001 horsepower at 6,000 rpm and 922 lb-ft of torque between 2,200 and 5,500 rpm.

Bugatti was famous for using engine-turned metal surfaces, and it does the same with the Veyron. From the top of the center console: the clock; vents; climate and sound system controls; and the shift lever for the amazing seven-speed, twin-clutch automatic transmission, which can be used in full automatic or shifted manually.

From behind, it's possible to see two important elements of the Veyron's well-defined aerodynamics package: the rear spoiler in the high handling position, where it can take an angle of 15–27 degrees, and, under the back of the Bugatti on either side of the exhaust opening, the venturis that help draw air from under the car.

The AMG SLS was the first car created and built by AMG, which was once just a subsidiary of Mercedes-Benz located away from Stuttgart. Here, race driver Tommy Kendall does his thing with the AMG SLS, powering through oversteer thanks to the brute force of the 6.2-liter dry-sump V-8.

Mercedes-Benz SLS AMG 2010

197^{mph}

197^mph

Mercedes-Benz's history with sports cars can be traced back to the 1920s and the famous SSK. Who can forget the generations of SLs or the SLR McLaren? Is the AMG SLS a worthy successor?

AMG was founded in Affalterbach, not far from Stuttgart, as an independent tuner of Mercedes-Benz cars. The automaker took over AMG as of 2005, and it is now the high-performance side of Mercedes.

The SLS was the first car designed and built by AMG. It was first shown at the 2009 Frankfurt Motor Show and on sale from 2010 to 2015 with two body styles, a roadster and a coupe, the latter with classic gullwing doors. The German supercar is based on an aluminum chassis/body frame. Match that with an aluminum body and the SLS tips the scales at just 3,573 pounds.

To motivate that light weight is a 6.2-liter, dry-sump, normally aspirated V-8 in three horsepower levels. It starts at 563 horsepower and 479 lb-ft of torque. Go with the GT model and that increases to 583 horsepower and 479 lb-ft, while the limited Black Series takes that to 622 horsepower and 468 lb-ft. In any case, the transmission is a rear-mounted seven-speed dual-clutch automatic, and the driveshaft is of lightweight carbon fiber.

The rarest of the SLS AMG models—around one hundred sold—has an electric drive, four motors that combine for 740 horsepower. And Mercedes and AMG couldn't resist racing the SLS, so there was a GT3 version, twenty-five built to meet the FIA's GT3 specifications.

Depending on which version of the SLS you have, you'll likely get to 60 miles per hour in 3.5–4.0 seconds and on to a top speed above 180 miles per hour. The fastest of the bunch—not counting the GT3 version—would be the Black Series topping at 197.

Some of us were lucky enough to test the SLS AMG GT at Willow Springs Raceway east of Los Angeles. The seats are contoured to keep you in place at speed, and straight ahead are a classic speedometer and tachometer. The center console is something of a sea of buttons, but on track all you are about is the one that selects Comfort, Sport, Sport Plus, or Manual control.

Odd as it sounds, the feeling one gets driving the SLS is one of being planted. You sense being securely grounded, the Mercedes feeling wide with its double A-arm/stabilizer bars suspensions front and back. Get to the end of the straightaway, on to the vented disc brakes, and turn smoothly into Turn 1.

Figure on a price of $150,000–$180,000 at auction.

Although production of the SLS AMG ended in 2015, it has a worthy successor. Called the Mercedes-Benz AMG GT, it is visually similar to the SLS AMG, though with an even more aggressive grille. Available as a coupe or roadster, both have conventional doors. AMG went from the non-supercharged 6.2-liter V-8 to a twin-turbo 4.0-liter V-8 with 469 or 550 horsepower. In either case, 0–60 comes up in less than 4.0 seconds through the seven-speed gearbox. Top speed is just shy of 200 miles per hour.

We can guarantee it's a worthy successor. We found out on a road on the east side of California's Napa Valley, winding past a lake and into the hills. Tree-lined, the road carves quickly left and right, and all you can do is grin as the AMG easily darts with it. May this road never end.

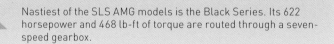

Nastiest of the SLS AMG models is the Black Series. Its 622 horsepower and 468 lb-ft of torque are routed through a seven-speed gearbox.

Rarest of the SLS AMG variations is the Electric Drive, with only one hundred sold. They were priced around $435,000.

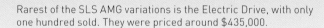

Thanks to the extensive use of aluminum in its chassis and body, the SLS AMG weighs in at an impressive 3,573 pounds. Depending on the version of the SLS and its V-8, you'll be at 60 miles per hour in 3.5–4.0 seconds as you power on to a top speed just shy of 200 miles per hour. Each AMG engine is built by one person who signs the engine. At 6.2 liters, the V-8 is rated at 563–622 horsepower, depending on model.

The body shape of the Huayra was created around what
Pagani calls active aerodynamics and is done in carbon fiber.

Pagani Huayra 2012

238^{mph}

When Horacio Pagani started his exotic car company in the early 1990s, we were amused . . . and curious. Was this young man to be the next Enzo Ferrari? Could he really make it in that exotic car zone near Modena that already featured time-honored Ferrari, Lamborghini, and Maserati?

Born in Argentina, Pagani knew he had to emigrate to Italy to become the engineer he hoped to be. He did go on to become chief engineer at Lamborghini and founded his own firm, Modena Design. But creating supercars?

Pagani had the backing of his fellow Argentine, world driving champion Juan Manuel Fangio, who gave him an inside connection with Mercedes-Benz. That was an important relationship that began with the Pagani Zonda and continues today. And a major reason for Pagani's success, which we witness each year at the Quail Motorsports Gathering on the Monterey Peninsula. The faithful bring their Paganis, now their Huayras, literally from around the world.

What they are shipping to the United States is the Huayra, and we need to cover the name first. Like how to pronounce it. Closest we can come is "Why-ra," though your best bet is to check out one of the YouTube videos that explains it. It's named for Huayra-tata, the father of wind in Quechua culture . . . so much more complicated than Enzo or Impala.

Paganis have been unusual-looking supercars—not as sensuous, perhaps, as Ferraris, but also arguably more purposeful. What you see in the Huayra is a shape in carbon fiber built around active aerodynamics. That form and the flaps inside the car are designed to create the most downforce and least drag for any given driving situation. Apparently, they can work front/rear or left/right to create maximum stability.

Not a bad idea when you're dealing with a 6.0-liter, twin-turbo AMG Mercedes V-12 just aft of your shoulders. This 60-degree engine is pumping out 720 horsepower and 738 lb-ft of torque or more, depending on the version of your Huayra. Paddle your way through the seven-speed sequential transmission to the rear-wheel drive, and the needle will whip past 60 miles per hour in around 2.8 seconds.

Lift the rear bodywork, and the engine is a visual treat. That compartment with the drivetrain and rear suspension is as beautifully detailed and finished as luxury car interiors—such a mechanical jewel. At the back is an exhaust system fashioned in titanium that winds rearward to Pagani's signature four exhaust outlets.

Then there's the interior. Some would call it the most beautiful in the world, the most bling for the big bucks. Others would term it overdone. It is, to say the least, loaded with dials, switches, and buttons. They are

everywhere, many of them shiny, it could be hard to keep your eyes on the road after dark. Dramatic to say the least.

There has to be special mention for the exposed shift linkage, which is a sculpture of beauty.

So far Huayra has been written in the singular, but there's more to it than the reported one hundred gullwing-door coupes that were made after the supercar launched at the 2011 Geneva Auto Salon. The Roadster came along at the same show in 2017 and, again, was limited to one hundred vehicles. Then there are the special-order versions. The Carbon and White editions. Those twenty hyper Huayra BCs with their revised aero and 745 horsepower. The Pearl version and the three Dinastias that were assembled. But our favorite, even if just for the name? The one-off Huayra Monza Lisa.

Now, if we just had a few million dollars to spend on an exotic car . . .

Among the special editions of the latest Pagani is the hyper Huayra BC. Just twenty were made with revised aero and 745 horsepower . . . and a very high price.

Pagani doesn't just bolt the Mercedes AMG twin-turbo V-12 in the back of the Huayra but styles it in the engine compartment, beautifully detailed.

So, do you think it looks marvelous? A bit glitzy? Stylish? Or overdone? It's rare to find someone who doesn't have a strong opinion about Pagani interiors.

Pagani Huayra 2012

Porsche launched the production version of the 918
hybrid supercar at the 2013 Frankfurt Auto Show.
Deliveries began that December.

Porsche 918
2013

211^{mph}

In many ways, Porsche car model numbers make a great deal of sense—a logical progression through seasons and models: the 904, 906, 907, etc. race cars, then the 911 sports car, and so on.

And then sometimes they make no sense at all. The 917 was the legendary Le Mans–winning race car in 1970 and 1971. The 918 was launched at the Geneva Motor Show in 2010. Just two model numbers apart, but forty years apart - so separate time wise and yet they share one very important trait: both Porches set the automotive world on its ear.

One was an out-and-out race car with a raucous flat-twelve engine and went on to win worldwide, from Le Mans to Can-Am. The other is also known for its powerplant, a gas/electric hybrid that caused *Car and Driver* to declare, "Hi-Po Silver! Porsche's 918 Spyder is the quickest production car yet."

Let's look closer at the latter.

Porsche had already been dabbling in hybrid powertrains. For the all-wheel-drive 918, they began with the basic thirty-two-valve V-8 block used in the company's successful RS Spyder race cars. Fitted with a dry sump and a flat, 180-degree crankshaft, the 4.6-liter engine spins out 608 horsepower and 390 lb-ft of torque.

That's just for openers, because bolted in between the engine and the seven-speed twin-clutch PDK transmission is a 154-horsepower motor that can act on its own, add its power to the V-8, or become a generator to feed electricity back into the battery pack.

Driving the front axle is a 127-horsepower motor that works until the car is at 163 miles per hour. Add all that up and it comes, by Porsche's reckoning, to 887 horsepower and 940 lb-ft of torque ("calculated on the crankshaft, complete system in 7th gear").

Supplementing all this is the requisite battery, a 6.8-kilowatt lithium-ion pack made up of 312 lithium-ion cells and situated behind the seats.

Working in unison, this powertrain will get the 918, by *Motor Trend*'s measure, to 60 miles per hour in 2.4 seconds, to 100 miles per hour in 5.1 seconds, and though the quarter-mile in 10.0 seconds at 145.2 miles per hour. Don't lift and the 918 will continue to 211 miles per hour. At the other end of the performance scale, if you put the 918 in pure electric mode, it's good for a range of 18 miles but up to 93 miles per hour getting there.

In E-Power mode, the powertrain alternately mixes electric and gas power for the best in economy-oriented driving. Likely not why you bought a 918, so next comes Sport Hybrid. Now the V-8 works all the time, the electric motors kicking in when needed for better performance. Race Hybrid bumps things up a notch, the PDK programmed for faster driving, the electric motors on max. As you might imagine, Hot Lap draws full power from all components, but as the name implies, that's good for a lap or two before the batteries are depleted.

Bringing all this to a halt are carbon ceramic brakes, 16.1 inches up front, 15.4 at the back.

All this power rides as low as Porsche could get it in a monocoque chassis of carbon-fiber-reinforced polymer, its two-piece targa body made of the same material. An example of how they worked to get that low center of gravity: the PDK gearbox is mounted upside down relative to its use in other Porsches.

At the front are double-wishbone suspensions, while the rear is a multilink design with an electro mechanical system that allows for 3 degrees of steering, in-phase at high speeds, counter-phase when slower.

Covering all this is a body that is so decidedly Porsche you could recognize it minus badging. Originally sold for $845,000, the 918 production run of, appropriately, 918 units was quickly sold. Want one today? Double that price.

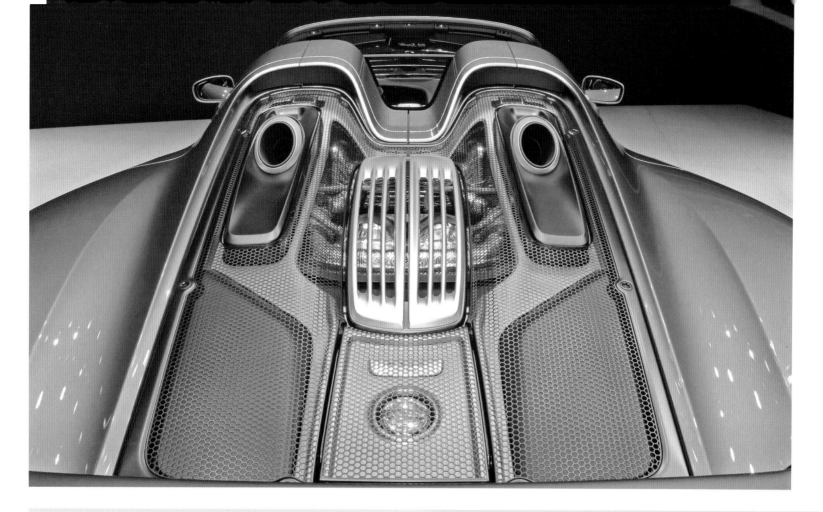

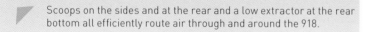 Scoops on the sides and at the rear and a low extractor at the rear bottom all efficiently route air through and around the 918.

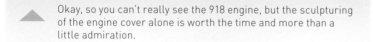 Okay, so you can't really see the 918 engine, but the sculpturing of the engine cover alone is worth the time and more than a little admiration.

 The first time many of us saw the 918 move was during the 2010 Monterey Car Week, when Porsche head of design Michael Mauer drove it for us.

Ahead of that blue flame lurks a hybrid drivetrain.
Match a 3.8-liter twin-turbo V-8 with an electric
motor for a total of 903 horsepower and 723 lb-ft of
torque. *GF Williams*

McLaren P1
2013

217 mph electronically limited

"McLaren" can mean many things. Bruce McLaren, the highly regarded New Zealand race driver who founded the company. Formula 1–, Indy 500–, Can Am– and 24 Hours of Le Mans–winning race cars. The legendary McLaren F1. Or any one of a dozen modern supercars.

Most of those British-built machines are in McLaren's Sports and Super Series. Each model is known by its Euro horsepower (PS versus brake horsepower), 540, 570, 650, 675, or 720, along with several special editions such as one named for race driver Ayrton Senna. Our subject is a member of McLaren's Ultimate Series, the P1.

If what you are about to read tickles your fancy, sorry, but you're too late. McLaren hand-built 375 P1s from 2013 to 2015, and it's possible they were all spoken for before the car was conceived. Sounds silly, but the P1 was meant to be the modern incarnation of McLaren's legendary F1, built from 1992 to 1998. Opening price for a P1 was $1.35 million

but, hey, why not add some options? Want one out of auction today? Put aside around $2.5 million.

As with LaFerrari and Porsche's 918, the P1 has a hybrid drivetrain. It starts with a 3.8-liter twin-turbo V-8 at 727 horsepower and 531 lb-ft of torque. Add the electric motor and the total is 903 horsepower and 723 lb-ft of torque. Among other things, the electric motor counteracts the turbo lag in the V-8. This prodigious power is routed to the rear wheels by way of a seven-speed dual-clutch automatic transmission.

The inner core of the P1 is a carbon-fiber monocoque—what McLaren calls a "MonoCage." Made of carbon fiber, it is the main piece of structure inside the supercar but also serves as a safety cell for driver and passenger. McLaren claims the material has " . . . more than five times the strength of top-grade titanium and twice the stiffness of steel." And yet it weighs just 198.4 pounds.

 Check out the fixed carbon-fiber seats. Forget power assist, but plan on easy-to-reach controls and nothing to distract you on the way to 60 mph in under 3.0 seconds. *GF Williams*

Chassis bits include double wishbone suspensions front and rear with, "independent hydro-pneumatic control" of springs and dampers. This all combines to produce predictable, flat cornering. The system can lower the body by 1.9 inches and bump up spring stiffness by 300 percent when the P1 is in Race mode. There are also Normal, Sport, and Track modes plus big ceramic brakes at each wheel and assisted rack and pinion steering.

A P1 weighs in at 3,075 pounds, and there isn't a pound of waste on the car. There's a decided athletic look to the McLaren, like the body was shrink wrapped around the structure. American Frank Stevenson, who designed the P1, explains, "I wanted to take out as much visual weight as possible . . . a car with absolutely no fat between the mechanicals and the skin. It's as though we stuck a tube inside and sucked all the air out—a dramatic, honest shape but also a very beautiful one."

That minimalist attitude carries over inside the P1. Fixed carbon-fiber seats and no power assist for anything. Controls all within easy reach and nothing to distract the driver. Good idea as she/he is likely busy getting to 60 miles per hour in less than 3.0 seconds, and through a quarter-mile in 9.7 seconds, on to that top speed of 217, electronically limited. It's said that if you figure out how to override the limiter, you could darn near hit 250 miles per hour.

Inside that snarly bodywork is a carbon fiber "MonoCage." It acts as both a structure for the supercar and a safety cell for driver and passenger. *GF Williams*

This isn't any old rear wing, but a decisive device that works with the double wishbone suspensions front and rear plus the "independent hydro-pneumatic control" of springs and dampers for fast, stable driving. *GF Williams*

Huracán, as in hurricane, has been a big seller for
Lamborghini, making up almost 70 percent of the Italian
automaker's sales. While some exotic auto companies
go for a rounded exterior shape, Lamborghini prefers a
harder edge and the use of the hexagon shape.

Lamborghini Huracán 2014

201^{mph}

201mph

You can think of Lamborghini's Huracán as an exotic car that comes in five flavors that range from just over $200,000 to shy of $280,000. Successor to the highly regarded Gallardo, of which Lambo sold 14,022, it is the main model in the Italian automaker's lineup. For instance, in 2016, Lamborghini sold 2,353 Huracáns out of a total of 3,457 sales that year.

What do those five flavors share?

Aluminum is used for both the chassis and the outer bodywork. Carbon fiber is also part of the chassis, as are composite materials in the body. The design of that body is decidedly Lamborghini: not the rounded, flowing shapes one finds on a Ferrari, but harder edged. The automaker likes to point out its use of the hexagon in the design's front air intakes, overall shape and mesh, side windows, and engine intakes.

This shape makes a great comparison with Ferrari's 488 GTB, each a distinct statement by its maker. As with the 488, there are both coupe and open-top spyder Huracáns.

That hexagon theme repeats in the interior of the Huracán in the shape of the instrument cluster, and air vent, and the pattern in the dashboard inset. Where Ferrari keeps most controls on the steering wheel or within small nearby clusters, the Huracán has just a few controls on the wheel, but a sea of buttons and switches fronts the top to bottom of its center console.

Double wishbone suspensions at both ends have steel springs and hydraulic dampers. The brakes are carbon ceramic.

That mid-mounted engine is a 5.2-liter V-10, another important variation from Ferrari. The gearbox is a seven-speed dual clutch automatic. And here's where we begin to get that separation in flavors.

Huracán 620-2s race in several series in the United States, from GT competition in the IMSA series to their own series, Lamborghini Super Trofeo.

Just as it should be in a supercar: a huge central tachometer with that prominent, hard-to-miss redline for the V-10.

Aluminum and composites are used to create the Huracán's chassis and outer body panels, and it is offered as both a coupe and a roadster.

Below the Huracán's taillights, you can again see the use of the hexagon shape in the lower grille work.

The main Huracán has 602 horsepower and 413 lb-ft of torque and a 0–60 time of 3.2 seconds. In the coupe and spyder—model LP 610-4—that power is sent to all four wheels. Flavors two and three—coupe and roadster—use the same V-10 and are called LP 580-2. Now the horsepower is 572 and sent just to the rear wheels. Slightly lighter than the four-wheel drive versions at 3,062 pounds, two-wheel drive versions had their weight set at 40 percent front/60 percent rear and had three driving modes: Strada, Sport, and Corsa. In Lamborghini's words, this was done to " . . . tune to proved oversteering characteristics," if that's your cup of tea.

So, what's the fifth flavor of Huracán? That would be the LP 640-4 Performante. As you can tell by the name, now the V-10 pumps out 631 horsepower and 443 lb-ft to all four wheels. The chassis has been retuned for added power and purpose with active aero and aero vectoring.

A Performante was rushed around the Nürburgring's Nordschleife in a record 6:52.01.

Ferruccio Lamborghini, didn't allow factory racing, but Lamborghini's current owner, Volkswagen AG, is all for it. They make nicely nasty looking racing cars—Huracán 620-2s in several forms raced around the world. You'll find them in racing in IMSA (International Motor Sports Association); in their own series, Lamborghini Super Trofeo; and in the GTD class against the likes of Ferrari, Audi, and Porsche in the United States.

And guess what? There's also a police car version of the Huracán, but that's a whole different story.

And the hexagon shape once again, now in design elements in the Huracán's center console and its sea of controls. Sporty, yet elegant.

While founder Ferruccio Lamborghini wasn't in favor of racing the company's cars, Lamborghini's current owner, Volkswagen AG, promotes them in international competition.

Beautiful as it may be, the exterior design of the 488 GTB was shaped to cheat the wind, aiming airflow for maximum downforce.

Ferrari 488 GTB 2016

205mph

If Enzo Ferrari were alive today he'd be around 120 years old, and he'd put his stamp of approval on the 488 GTB. Truth be told, there were several Ferraris that, in their day (the 1980s), were just okay. Nothing special.

It was Luca di Montezemolo and Amedeo Felisa who brought Ferrari out of the doldrums in the 1990s, and while both are no longer with the automaker, the 488 GTB (as in Gran Turismo Berlinetta) bears their stamp.

This time we start in the driver's seat. No surprise—it's all about driving. The flat-bottom steering wheel is fitted with all the most immediate needs, from starting the car to turn signals to windshield wipers to shock and suspension settings. Want to change from Race to Sport to Wet mode on the *manettino*? It's all right there.

It's not as comprehensive as a Formula 1 steering wheel, but just as involving given what you'd do on the open road. Within finger-tip reach left and right are most all other functions, such as the audio system. Only the light switch on the lower left and the heating/air conditioning controls aren't on the wheel . . . and they aren't far off.

The point, of course, is that this is a driver's car, designed for just that and not as a study in cute ergonomics, as some supercars seem to be.

Just back of the driver's seat is a major reason for all this seriousness.

Predecessor to the 488, the 458 Italia was powered by a normally aspirated 4.5-liter V-8 that offered 562 horsepower and 398 lb-ft of torque. For the 488, Ferrari went the turbo route, using a dry-sump V-8 with a displacement of 3.9 liters and adding a pair of twin-scroll turbochargers, each with its own intercooler. Output increased to 661 horsepower and 561 lb-ft of torque. Both cars use a seven-speed Getrag dual-clutch automatic gearbox.

The suspension is a double wishbone front/multi-link rear on an aluminum chassis that is covered with a body of the same metal. The 458 Italia was already a visual heartbreaker, but the 488 GTB takes it another step. It is so, well, *Ferrari* looking—lithe, sweeping lines that are both beautiful and purposeful. The so-called "aero pillar" in the front center of the grille moves some air under the car and to the radiators on both sides. That rear open channel behind the glass engine cover flows air to create downforce. So too do the hidden forms under the 488.

No shape on the body is there just for looks, but they sure look good. And then there are the two scoops on the front of the rear fenders. What's cool is that you see them in the rearview mirrors, and they sort of hint at the past—1964 250 LM race car? —but their purpose is very modern: to flow air to the intercoolers and to reduce drag.

Happily, the 488 comes in two forms: the GTB coupe and the open-top spyder, and with either case your bank account will diminish by a quarter million dollars or more. While it used to be that a convertible version of a closed sports car often suffered for its less rigid body, Ferrari claims the spyder has 95 percent of the coupe's rigidity and has the same 0–60 time (3.0 seconds) as the coupe, despite its added 110 pounds.

It was a coupe we had at Spring Mountain raceway along with several other exotic cars. Cutting hot laps, all were fun, of course, but there was something special about the 488. It felt even more in touch with the circuit, more ready to work with you, easy to place, plenty of feedback . . . and the most fun to drive.

Enzo Ferrari would be proud.

 Isn't it wonderful what modern technology has allowed designers to do with headlights? A small main beam down low with a stack of accessory lights.

Under that rear deck lid is a 3.9-liter dry-sump V-8 with two turbos, 661 horsepower and 561 lb-ft of torque, plus a seven-speed dual-clutch transmission.

At the 2016 Geneva Auto Show, Bugatti unveiled a version called Veyron Sport Vitesse, which bumped horsepower from the W-16 to 1,200.

Bugatti Chiron
2016

261 mph (limited for safety reasons)

This story could be subtitled, "Theme and Variations." The song begins in 2005 with Bugatti's Veyron, which evolved into the Grand Sport (2009), Super Sport (2010), and Grand Sport Vitesse (2012). Visually, they didn't change much, the big bump coming out back as the four-turbo, 8-liter W-16 grew from a mere 987 horsepower to 1,200.

For those of us who drove the first version on the roads around the factory in Molsheim, Alsace, in France, and later had the chance to top 200 miles per hour on a banked track with the Vitesse, it's been quite a journey. There's something about those blown sixteen cylinders humming just behind you that's magic.

As of March 2016 and the Geneva Motor Show, the Veyron and its iterations became old hat. Bugatti uncovered the Chiron, which can be compared to the earlier Bugattis by the old phrase, "The same, only different."

Much of that difference, visually, comes from the Bugatti Vision Gran Turismo. Many automakers did the same, penning an outrageous variation of a production car that could be raced on-screen in the Gran Turismo video game. In fact, Bugatti built one example of the car and sold it to a private owner for some tens of millions of dollars.

Many of us saw this one-off at the 2016 Pebble Beach Concours d'Elegance and recall what had been toned down from the video game version for the reality of the Chiron. Most in your face is the large "C" on the carbon-fiber body's side. It starts with the line that runs rearward and sweeps up on the Veyron, but now it continues up and over the side windows. Like the Vision's, the Chiron's headlamps are inset. The tail lamps run horizontally across the back of the car.

The same, only different—but an exciting difference.

That "C" theme continues inside the Chiron to create a decidedly driver/ passenger cockpit. The passenger has little to do but sit there while the driver is easily entertained with a flat-bottomed steering wheel, and a well-done mix of colors and chrome. Gauges straight head, tach on the left, navi map on the right, and dominating it all? A 500-kilometer-per-hour speedometer. That would be 310 miles per hour.

And given enough highway, courage, and the Chiron's potential, you could wind that speedo up to the 420-kilometer-per-hour (261-mile-per-hour) mark.

While it maintains many of the Veyron/Vitesse tech specs, the turbo 8.0-liter W-16 has been though a major upgrade. How about to 1,479 horsepower and 1,180 lb-ft of torque? With its Ricardo seven-speed dual-clutch automatic and all-wheel drive, that's enough to scoot the

Seen at the 2017 Cannes Film Festival, the big "C" on the side of the Bugatti Chiron honors famed Monegasque race driver Louis Chiron. *Shutterstock*

The first of the modern-day Bugattis came in 2005 in the form of the Veyron with its unique design and an 8.0-liter, four-turbo W-16 with around 1,000 horsepower.

Chiron owners exist in rarefied supercar air, and the colors of their Bugattis often reflect that. *Shutterstock*

Chiron to 62 miles per hour in 2.4 seconds, to 124 miles per hour in 6.1 seconds, and to 186 miles per hour in 13.1 seconds.

All this works through a carbon-fiber tub and a chassis that maintains the Veyron/Vitesse 106.6-inch wheelbase and double wishbone suspensions front and rear with electronically controlled shocks that can vary the ride height.

Price? That can depend on the dollar/Euro exchange rate, but you would be at $2.5 million plus.

Oh, and the "C" theme and the name? That's to honor Monegasque driver Louis Chiron, who won such varied events as the 1931 Monaco Grand prix (his home race), the 1933 Spa 24 Hours with Luigi Chinetti, and the 1954 Monte Carlo Rally. A man worth honoring.

Here, the DB11 can be seen in all its front-engine, exotic car glory, with its large grille and low roofline. Depending on the engine, 60 miles per hour comes up in 3.8 seconds with the V-12 or 4.0 seconds thanks to the V-8. *Courtesy Aston Martin.*

Aston Martin DB11 2016

200^{mph}

Run through the list of exotic cars in this book and you'll find most of them have a sugar daddy. Ferrari is part of Fiat Chrysler Automobiles. Bugatti and Lamborghini are under Volkswagen's umbrella. AMG is Mercedes and Acura is Honda. And Aston Martin is different.

Founded in 1913, it has passed many times from one private owner to another. Not big automakers, but individuals or small groups. There was a twelve-year period under Ford, but then it was back to private funding. There is now a 5 percent hook-up with Daimler Benz, but that's minor in the ultra pricey world of exotic cars.

None of this has stopped Aston from creating some very exciting cars over the decades: James Bond's DB5. Cars with names like Vantage and Vanquish and Virage. Many with the surname DB, the initials standing for one-time owner David Brown.

Today those initials are followed by 11, as in DB11, the latest in the line. Most Astons have been exciting to look at from the DB4GT Zagato on. There was a design theme set by Ian Callum with the 1993 DB7 that you can still see hints of in Marek Reichman's design of the DB11. It's a stunner: that classic Aston grille shape, the low roof flowing back to the rear spoiler. The DB11 looks like it's moving when it's standing still—arguably the epitome of an front-engine exotic car. It's hard to say if it looks better as a coupe or a roadster (the Volante) with its multi-layer top folded.

There's a mix of sport and elegance with the interior. Super stitching on the seats. Most controls are centered in the middle console, with some of the electronics, such as the infotainment/navigation system, from Mercedes-Benz.

Double wishbone coil spring suspensions are fitted fore and aft and have three possible modes: Normal, Sport and Sport Plus. And, of course, there are four-wheel disc brakes.

It would be hard enough to order a DB11, choosing off the option list, but first you'd have to decided on the engine. There are two for its rear drive. Most intriguing is the twin-turbo 5.2-liter V-12. That provides 600 horsepower and 516 lb-ft of torque through a rear-mounted ZF automatic transmission. Figure on 0–60 in 3.8 seconds and on to that 200-miles-per-hour top speed.

An alternative is provided by Mercedes-AMG, a twin-turbo 4.0-liter V-8, similar to that in the Mercedes-AMG C63 S. Now you have 503 horsepower, but only 18 fewer lb-ft at 498 lb-ft. As a result, and using the same ZF gearbox, this version gets to 60 in 4.0 seconds. That's the blink of an eye compared to the V-12.

The V-8 version is just short of $200,000, while the virtually identical V-12 DB11 starts at more than $216,000. But that's the funny thing about many buyers of exotic cars. Those who watch their pennies would wonder at spending the extra dollars for the V-12. Those who sell these machines will tell you about customers who just want the most expensive model. Period.

There is an under-hood choice of engines for the DB11. One is a 5.2-liter, twin-turbo Aston V-12 with 600 horsepower. The other is a Mercedes-AMG twin-turbo V-8 at a mere 503 horsepower. Both will get you where you need to go very, very quickly. *Courtesy Aston Martin.*

Check out the stitching on those leather seats. The big speedo sits straight ahead, and the center console is filled with a screen and switches. But the most important control? That gas pedal down low and to the right in the driver's foot well. Depress with care. *Courtesy Aston Martin.*

As slick and sleek as it looks, today's Ford GT's exterior design is all business—aero business—to help the car slip through the air and to stay planted to the ground. Major signatures of the shape are the dramatic flying buttresses ahead of the rear fenders. *Kevin McCauley*

Ford GT
2017

216^{mph}

We've now had four generations of Ford GTs. There was the original GT40 in small- and big-block form in the early 1960s. GT40 Mk IVs won Sebring and Le Mans in 1967. The 2005–2006 Ford GT that borrowed much visually from the originals. Now a new Ford GT that doesn't borrow much from the past, but for its name.

We first saw the new Ford GT at the 2015 Detroit Auto Show and, to say the least, it is dramatic. Thanks to carbon-fiber bodywork, Ford designers were able to give a hint of Ford GT heritage in the nose and then a sweep back to a killer flying buttress that would have been very difficult to form in aluminum of the sort used on the 2005 GT.

As dramatic and cool as it looks, none of that shape is there for fun and appearance. It all works, flowing air to all the right places for cooling and downforce. At the back is a deployable wing that works in unison with flaps inside large ducts up front. When the wing rises at speed to provide more downforce, the ducts close to balance that force front-to-rear. Lower the wing and the flaps open.

All this bodywork wraps around a monocoque of carbon fiber that has an integrated roll cage of steel and front and rear aluminum subframes.

Front and rear suspensions have upper and lower wishbones with—à la many race cars—pushrods and rocker arms working inboard springing, all the better for unsprung weight and packaging. That springing is a combo of coil springs and torsion bars.

As is common in supercars, the Ford GT offers multiple driving modes via a steering wheel control: Normal for daily driving; Wet in the slick stuff; Sport to kick it up a notch; Track for, well, on the track; and V-Max

for maximum straight-line speed. Not surprisingly, they alter such things as suspension stiffness, traction and stability control, transmission tuning, and throttle calibration.

What's cool about the Ford GT is its ride height adjustment. In Track and V-Max modes, the suspension lowers almost 2 inches. With Track, the rear wing rises and the forward ducts close for max downforce. Go with V-Max and the wing lowers for best aero and top speed. At very low speeds, you can also raise the nose a bit to keep from spoiling the expensive front spoiler on obstacles.

So what gets the slick aero 3,054-pound Ford GT to 60 miles per hour in less than three seconds and on to 216 miles per hour?

A non-hybrid EcoBoost V-6; an aluminum, dry-sump 3.5-liter; twin turbo with dual cylinder-and-port injection; 647 horsepower; and 550 lb-ft of torque. All this spins rearward through a Getrag seven-speed, dual-clutch transaxle. Just in case, Brembo disc brakes are in the front and rear.

The driver gets to view this from high side support seats, behind a flat-bottom steering wheel with fifteen major switches and, ahead, a digital dash that changes configuration with each drive mode. All controls very logical and easily within reach.

This all-American Ford GT, painted red, white, and blue, is assembled in Canada. That is a good thing, because all the work is done by Multimatic in Markham, Ontario, a highly regarded automotive engineering company with a solid background in racing.

Very important given Ford's desire to race the newest GT. In fact, a GT won its class—LM GTE Pro—in the 2016 24 Hours of Le Mans, fifty years after a Ford GT40 MKII took the overall win in the French classic. Chip Ganassi Racing puts the Ford GTs on-track throughout the world, where it is always nip and tuck with Ferrari, Porsche, and Aston Martin.

Good to have a Ford GT back in the mix, and for $400,000 you can join the club.

Inside that sculpted Ford GT body is a monocoque of carbon fiber with a steel roll cage and front and rear subframes done in aluminum. That rear wing rises to increase downforce and stabilize the supercar at speed. *Kevin McCauley*

Inside that nose are flaps that close at speed as the rear wing deploys. Lower the wing and the flaps open—all the better for stability at speed. And how about those well-fashioned headlamps? *Kevin McCauley*

Note the non-round steering wheel behind the digital instrument panel. The latter changes its configuration when the driver opts between the various driving modes such as Normal, Sport, or Track. *Kevin McCauley*

Lexus puts the LC 500's 0–60 time at 4.4 seconds, which feels about right given the luxury level of the two-passenger coupe. Handling of the LC 500 is nicely balanced, giving you a sense of security as you tuck its nose into a corner. Responsive steering and brakes match the overall performance.

Lexus LC 500
2017

168mph electronically limited

Lexus has long meant luxury, but not necessarily excitement—unless quiet and solitude tickle your fancy. The manufacturer tried exotic excitement once before with the LFA. Developed from 2011 to 2016, it had good creds. How about a 520-horsepower 4.8-liter V-10, a six-speed sequential gearbox, and 0–60 in 3.9 seconds? How about a base price of $375,000?

Nice try, Lexus, but it just didn't take, and the automaker retreated to its luxo sedans, SUVs, and a few coupes. Turns out, however, that Akio Toyoda (yes, it is spelled differently), who heads the automaker, wanted to up the image of Lexus, so their engineers got back to it. The results are the LC 500 and LC 500h hybrid, and they were well worth the work. And so different from the LFA.

Two engines were offered: a 5.0-liter V-8 with 471 horsepower or a hybrid powertrain using a 3.5-liter V-6 and a pair of motors.

Let's start with the visuals. Some of us aren't exactly in love with Lexus's spindle grille, but it works for the LC 500s. It isn't as large in the overall design, and the texture works elegantly. There's something about the stance of the LC 500s—aggressive haunches at the back and a sense that the car is about to leap forward. Cool taillights and a coefficient of drag of 0.33.

The interior is downright scrumptious. Okay, an odd word for the inside of an automobile, but it works. Very comfortable seats, central gauge straight ahead, a large hooded info screen, all very driver oriented. Nice place to sit even if you weren't driving.

But that was the point of being at Texas Motor Speedway and its road course. Lexus puts the LC 500's 0–60 time at 4.4 seconds, and when you're in such a luxurious sports car, that feels about right. The car is neither small (187.4 inches long) nor light (4,280 lbs), and, in standard

Lexus managed to give the interior of the LC 500 a look that is stylish and yet sporty, as you can see from the center console grip handle for the passenger.

Lexus's earlier shot at doing an exotic car was the 2010 LFA with 520 horsepower. It could reach 0–60 in 3.9 seconds and had a price of $375,000.

Aggressive haunches frame the LC 500's rear and an overall shape that comes in with a coefficient of drag of 0.33. Though Lexus added a slightly formal look to the LC 500, the taillights and aggressive rear demeanor suggest something more . . . feral.

Note the large tach straight ahead, a good-sized infotainment screen properly hooded against the sun, and an elegant simplicity to the interior layout.

LC 500 form, comes with a 5.0-liter, thirty-two-valve V-8 with 471 horsepower and 398 lb-ft of torque. The transmission is a ten-speed paddle-shift automatic.

Alternatively, there's the LC 500h as a hybrid. This system matches an Atkinson-cycle 3.5-liter V-6 with a pair of electric motor/generators that combine for 354 horsepower. There is also a pair of transmissions: a CVT and a four-speed automatic. In concert, they provide what is, in a sense and in M mode, a ten-speed transmission.

It's odd to think of a buyer looking for a fast, fine-handling super coupe heading for a Lexus dealer. With the advent of the LC500, looking for such a car from Lexus wouldn't be a mistake. Akio Toyoda's aim was to put Lexus on the performance car map, and he seems to be on target. Prices start just under $95,000.

Looks like the generation-four ZR1 could take a bite out of you, doesn't it? Those mighty maws combine to cool the 775-horsepower supercharged V-8 and aid in aerodynamic stability. Both are important aspects considering the ZR1's 0-60 time of less than 3.0 seconds and top speed of 212 miles per hour. *Richard Prince*

Corvette ZR1 2019

212^{mph}

A brief history lesson.

Chevrolet's 2019 Corvette ZR1 is the fourth model to carry that now-legendary designation. First launched in 1970, the Stingray ZR1 was meant for serious drivers and racers. Its 350-cubic-inch V-8 had 50/370 horsepower, the spec sheet was competition driven—the "Rock Crusher" gearbox—and the option list limited, as was production. Just fifty-three were built in 1970–1972—rare birds now worth $200,000 and up.

Come 1990, the Gen II ZR1 boasted a 375-horsepower, Lotus-designed, Mercury Marine–built four-cam V-8 that is arguably the coolest-looking engine in General Motors' history. The 1996 C6 ZR1 featured a supercharger atop the 638-horsepower 6.2-liter V-8 and a polycarbonate window in the hood to view it.

Now we have the Corvette C7 version, and it ups the ante again. As with any ZR1, we'll start with the engine. Chevy stuck with a big-displacement—6.2 liters—pushrod V-8. On top goes a 2.65-liter supercharger, making it 65 percent larger than the blower on the C7 ZO6 'Vette. Among its slick equipment is a dual fuel injection system . . . one injector for the cylinder, one for the intake manifold.

The result is 775 horsepower and 715 lb-ft of torque. You can opt for a seven-speed manual or an eight-speed paddle-shifted automatic transmission. To cool the engine, gearbox, and electronic differential, the ZR1 has a total of thirteen radiators.

Plus, four levels of sound. You can sneak around town quietly in Stealth mode, pick up the volume through Tour and Sport modes, or blast through with Track.

What you really want to know is, as we used to say, "What will she do?" Less than 3.0 seconds to 60 miles per hour, through the quarter-mile in less than 11 seconds on the way to a top speed of 212 miles per hour.

And you'll look good doing it. Out front is a mouth of a grille that gulps air for that big motor and has an underwing to aid downforce. This is matched in back by one of two wings. One is somewhat conventional, riding along the upper edge of the rear of the deck lid. The other is downright nasty, an adjustable wing mounted up off the rear deck, à la race cars. It can provide up to 950 pounds of downforce, and while it trims top speed a bit, you'll still top 200.

The generation-three Corvette ZR1 was the first to get a supercharger, which huffed the 6.2-liter V-8 to 638 horsepower. You could see the blown engine through a polycarbonate window in the hood. What separates this and the fourth generation ZR1 from normal Corvettes is an increased sense and look of speed and purpose.

Likely the coolest-looking engine in General Motors' history is the 4-cam V-8 from the second-generation ZR1s. Designed in England by Lotus—which GM owned at the time—the aluminum engines were built by Mercury Marine.

The gen-two Corvette ZR1 disappointed some fans because while it had a somewhat different convex rear end than the standard 'Vette, the overall design difference wasn't all that much. Within a year all Corvettes had that back end anyway, so the main thing that separated the ZR1 was its 0-60 mph in 4.4 seconds.

For the first time since the original ZR1, this newest version will be offered as a coupe or convertible. There was a time when being a ragtop was a drag, the open body being more structurally loosey-goosey than the coupe. Not with the Corvette's aluminum chassis. The convertible's structure is identical to the closed version, except the well in which the top folds, and that adds fewer than 60 pounds. You can drop the top remotely or at any speed up to 30 miles per hour.

And you'll drop $119,995 for the ZR1 Coupe or $123,995 for the convertible. Pricey for a 'Vette, perhaps, but a bargain compared to much of the supercar competition.

INDEX

Abbey Panels, 91
Acura, 120, 122–123, 223
Aerospatiale, 111
Agassi, Andre, 117
Agnelli, Giovanni, 11
Alesi, Jean, 85
Alfa Romeo Carabo, 8, 12, 40, 116
Alfieri, 27
Allen, Paul, 100
Altieri, Giulio, 24
America Camoradi Scuderia, 162–163
AMG, 223
Andretti, Mario, 163
Artioli, Romano, 111
Aston Martin, 163, 228
    DB-7, 86
    DB11, 222–225
Attwood, Dickie, 167
Audi, 59, 94, 104, 118
    Quattro, 95–96
    TT, 21

Bangle, Chris, 41–43
Baron, Richard, 155
Bauer, 58–59
Bentley, 95, 186, 190
Bertone, 11, 14, 32, 40
    Stratos, 41
Bertone, Nuccio, 12, 14–15
Bilstein shocks, 91
Birmingham Show, 89

Bischof, Klaus, 54
Bizzarrini, Giotto, 14, 104–105
Blistein, 135
BMW, 127, 135
    M1, 56–61
    Motorsports, 133
    Procar series, 59
Bosch, 37, 159
Bose, 174
Bott, Helmuth, 96
Brabham, 59
    BT42-Ford, 128
    BT46B, 132
Bracq, Paul, 58, 60–61
BRDC (British Racing Drivers' Club)
    GT Challenge, 92–93
Brembo, 98, 122, 135, 141, 153, 164, 181–183
Bridgestone, 91, 156
Brown, David, 223
Brown, Lee, 116
BRP GT Championship, 136
Bryant, Tom, 47
Bugatti, 24, 136, 223
    Chiron, 189, 218–221
    EB110, 42, 110–113
    EB118, 185
    EB218, 185
    Type 57, 189
    Veyron, 112, 131, 161, 184–191, 219
    Veyron Sport Vitesse, 218

    Vision Gran Turismo, 219
Bugatti, Ettore, 186, 190

Callum, Ian, 223
Canepa, Bruce, 100
Cannes Film Festival, 221
Carrozzeria Bertone, 12
Casner, "Lucky," 162–163
Chandler, Otis, 100
Chevrolet, 116–117, 135
    Corvette, 115, 149, 173
    Corvette C6 ZR1, 235
    Corvette C7 Z06, 235
    Corvette Z06, 152
    Corvette ZR1, 128, 234–237
    Stingray ZR1, 235
Chinetti, Luigi, 20, 37, 221
Chip Ganassi Racing, 228
Chiron, Louis, 189, 221
Christensen, Michelle, 120–121
Chrysler, 21, 102, 104, 106, 108, 118
Citroën, 24–25, 27
    SM, 26–27
Cizeta Moroder V16T, 106
Clinton, Bill, 100
Connelly, 127
Cooper, John, 29
Cortese, Franco, 78
Coward, David, 59
Cresto, Sergio, 95
Cromodora wheels, 33
Cunningham, Briggs, 177

Daimler Benz, 223
Dallara, Giampaolo, 12, 59–61
Dayton, Bill, 107
De Dion, 27, 116, 118–119
De Tomaso, Alessandro, 14–15, 103
Dean, Warren, 100
DeLorean, 58
Dennis, Ron, 128, 136–137
Detroit Auto Show, 176–177, 227
Donington Park Grand Prix Circuit, 90–91
Donner, Bob, 34–35
Dreyfus, Rene, 186
Duntov, Zora, 115

Egan, Peter, 142–143
European BPR Championship, 85

Fangio, Juan Manuel, 24, 147, 197
Federation Internationale de l'Automobile (FIA), 65, 95–96, 165
Felisa, Amedeo, 215
Ferrari, 24, 40, 87, 94–95, 103, 121, 186, 191, 209, 228
    250 GTO, 65–66, 189
    280 GTO, 96
    288 GTO, 64–69, 86
    288 GTO Evoluzione, 82
    308, 12
    308 GTB, 32
    308GT/4, 40
    360 Modena, 181–182

488 GTB, 214–217
612 Superamerica, 32
Berlinetta Boxer, 21, 32–35, 40, 74
Boxer, 12, 24, 27–37, 72
Daytona, 10–21
Dino, 12, 32
Enzo, 121, 138, 143, 154–161,
  164–165, 173
Evoluzione, 78
F40, 69, 76–85, 91, 111
F50, 138–143
GTO Evoluzione, 69
Maranello, 74–75
Testarossa, 30, 37, 70–75, 131, 163
Ferrari, Enzo, 11, 17, 29–30, 77–78,
215
Fiat, 11
X1/9, 40
Fiat Chrysler, 223
Fioravanti, Leonardo, 12, 15, 17, 19,
21, 32, 78, 80
Ford, 21, 33, 152–153
  Falcon, 178
  Galaxy, 178
  GT, 176–183, 226–229
  GT40, 15, 131, 177–178
  RS200, 95
Ford, Henry, II, 176–177
Foyt, A. J., 177
Frank, Phil, 148, 153
Frankfurt Motor Show, 95–96, 185–
186, 189, 193, 200
Frere, Paul, 52, 95, 144–145

Gaffka, Doug, 179, 181
Gandini, Marcello, 8–9, 12, 14, 32,
  40–41, 102, 104, 106, 111–112

Garrett, 100
Gates, Bill, 100
Gendebien, Olivier, 30
General Motors, 33, 116, 188
Geneva Auto Salon, 198
Geneva Motor Show, 12, 24, 32, 40,
  146, 185–186, 201, 218–219
Ghia, 11
Ginther, Richie, 177
Giugiaro, Fabrizio, 189
Giugiaro, Giorgetto, 12, 14–15, 21,
  24–25, 27, 58, 60–61, 164, 185, 189
Goodyear, 116, 135, 181
Gregg, Peter, 50
Gregory, Masten, 177
Gurney, Dan, 177

Hannemann, Neil, 179
Helfet, Keith, 62–63, 89
Hermann, Hans, 167
Hilbig, Ditmar, 185, 190
Hill, Phil, 30, 52, 71, 75, 111–112, 139,
  142–143, 155, 163, 177, 191
Holbert, Al, 100
Honda, 135, 223
  NSX, 120–123
Hong, Patrick, 155, 159, 166, 174
Hull, Nick, 89

Iacocca, Lee, 102
IMSA (International Motor Sports
  Association), 59, 61, 85, 212
Indianapolis 500, 25
Indianapolis Motor Speedway, 24
International Race of Champions
  (IROC), 59
Italdesign, 11, 58

Jaguar, 94–95, 190
  D-type, 62–63, 87, 93
  E-Type, 11
  XJ13, 87
  XJ220, 62–63, 65–67, 86–93, 96,
    111
  XJ220C, 92–93
  XK120, 45, 87, 178
  XKE, 45, 178
John Paul II, Pope, 161
Jones, Davy, 87, 90, 93

Kendall, Tommy, 192
Kenwood, 132
Klaus, Ted, 121
Koni shocks, 82
Kott, Dog, 117
Kugelfischer-Bosch, 59

Laguna Seca, 85, 174
Lamborghini, 87, 186, 223
  350 GTV, 104–105
  Countach, 8–9, 12, 24, 32–33, 35,
    37–47, 59, 74, 102–104, 107–108,
    131, 173, 178
  Diablo, 91, 102–109, 185
  Gallardo, 181, 209
  Huracán, 208–213
  Miura, 10–21, 29, 45, 59, 131, 178
  Murcielago, 108
  P400, 12
  Super Trofeo, 212
Lamborghini, Ferruccio, 11, 32, 40,
  103, 212
Lancia, 94
  Delta S4, 95
  Stratos, 12, 40

Larson, Grant, 21, 172
Lauren, Ralph, 100, 131, 136–137
Le Mans, 20, 30, 34–35, 52, 59, 61, 85,
  92–93, 100, 128, 136, 152, 163, 165,
  167–168, 171, 176–178, 182, 201,
  205, 227–228
Leno, Jay, 18, 45, 98, 131, 173, 178, 188
Lexus LC500, 230–233
Lola T70 Coupe, 41
Los Angeles Auto Expo, 116
Losee, Richard, 155, 160
Lotus, 163
  Elan, 188
  Esprit, 58
Lutz, Bob, 106

Mangusta, 15
Maranello, 30
Marchesi, 59
Martin, Dean, 18
Maserati, 103
  BCTF Boyle Special, 25
  Bora, 22–27
  Ghibli, 27
  Khamsin, 40
  MC12, 162–165
  Merak, 26–27
Materazzi, Nicola, 78
Mauer, Michael, 203
Mays, J, 179, 181
Mazda Miata, 131
McLaren, 118
  F1, 114, 126–137, 160, 173, 186,
    188, 191, 205
  M6, 128
  M6GT, 128–129
  P1, 124–125, 204–207

McLaren, Bruce, 128, 205
Megatech, 118
Mercedes-AMG, 224
Mercedes-Benz, 116, 136, 145–146,
    192–195, 198–199, 223–224
    SLR, 131
Michelin, 82, 153, 172, 186–187, 189
Michelotto, 85
Mille Miglia, 144
Miller, Kenper, 59
Millibrook, 90–91
Mimran brothers, 103
MIRA, 90–91
Mitchell, Bill, 115
Modena Design, 197
Monaco Grand Prix, 127, 221
Monte Carlo Rally, 221
Monte Carlo Sporting Club, 127
Monterey Car Week, 203
Montezemolo, Luca di, 121, 158, 215
Mulsanne Straight, 92, 152, 167
Multimatic, 228
Munari, Sandro, 39
Murray, Gordon, 118, 126, 128, 131–136

Nardi, 127
National Association for Stock Car
    Racing (NASCAR), 44, 47
    173, 167
New York Auto Show, 118
Nielson, John, 91
Nürburgring, 90–91, 212

Ojjeh, Mansour, 128
Okuyama, Ken, 121, 156, 158
Orsi family, 24
OZ Racing, 135

Pagani
    Huayra, 196–199
    Zonda, 144–147, 197
Pagani, Horace, 145–147, 197
Pardo, Camila, 179
Paris Auto Show, 8, 17, 59, 167,
    185–186
Paris-Dakar Rally, 100
Pebble Beach Concours d'Elegance,
    219
Peugeot, 94
    T16, 95
Piech, Ferdinand, 186–187
Pininfarina, 11, 15, 17, 21, 30–32,
    34–37, 66, 69, 74, 78, 80–82, 139,
    156, 158
Pininfarina, Sergio, 15, 158
Pirelli, 59
Pontiac, 65
Porsche, 11, 57, 59, 61, 116, 136, 228
    911, 49, 53–54, 98
    917, 41, 115, 163, 186
    918, 200–203
    959, 35, 65–67, 69, 80–81, 85–86,
        89, 94–101, 111
    Boxster, 21, 172
    Carrera GT, 21, 166–175
    Cayenne, 167, 172, 174
    Stability Management, 53
    Turbos, 48–55
Porsche, Ferdinand, 186
Porsche-Steuer-Kupplung, 96, 99
Posey, Sam, 37
Project 132, 102

Quail Motorsports Gathering, 197
Queener, Chuck, 24

Randall, Jim, 89, 135
Raphanel, Pierre-Henri, 136
Razelli, Giovanni, 78
Recaro, 60
Redman, Brian, 163
Reichman, Marek, 223
Reynolds, Kim, 118
Riccardo, 187
Ruf, Louis, 152

Saleen, 163
    S7, 148–153
Saleen, Steve, 148, 152–153, 177, 179
Santoro, Michael, 118
Scaglionie, Franco, 104–105
Schultz, Peter, 53
Schumacher, Michael, 111, 155–156,
    159
Sebring, 149, 227
Seinfeld, Jerry, 98, 100
Senna, Ayrton, 205
Shaw, Wilbur, 24–25
Shelby, Carroll, 177
Shirley, Jon, 181–182
Spa 24 Hours, 221
Speedline, 91, 141
Stanzani, Paolo, 12, 42, 111–112
Stephenson, Frank, 164, 206
Stevens, Peter, 114, 118, 132
Suharto, "Tommy," 104, 118

Texas Motor Speedway, 232
Thomas, Freeman, 21
Toivonen, Henri, 95
Tokyo Auto Show, 186
Tom Walkinshaw Racing (TWR), 89
Tour de Corse, 95

Toyoda, Akio, 233
Transportation Research Center, 35,
    37, 49
Trasformazione Italiana Resina, 59
Turin Motor Show, 12, 32

Uehara, Shigeru, 121

Vector, 114–119
    Avtech Roadster, 119
    Avtech WX3, 118–119
    M12, 114, 118
    S/C, 119
    W2, 115–116, 119
    W8, 116–117
    W8 Twin Turbo, 119
Volkswagen, 59, 111–112, 185–186,
    212, 223

Walkinshaw, Tom, 89, 92
Wallace, Bob, 12
Wankel, 116
Warhol, Andy, 59, 61
Weber carburetors, 27, 32, 35
Weber-Marelli, 66, 82, 85
Wiegart, Gerald "Jerry," 116–119
Wolfkill, Kim, 164–165
World Driver's Championship, 30

Zampolli, Claudio, 106